Glenological Chemistry

The Organisation of Matter

The Periodic Table of Elements

as an expression of the Atkinson Conjecture

*"You have my vote for making sense of the periodic table. I'm always following
up and confirming your intellectual realizations. My hat's off to you, my friend.
I owe you a great (and growing) debt for deeper understanding of biodynamics
and beyond, which I hope including mention of your work in the tail
end of my book, Quantum Agriculture, is a first payment for.
Thank you, thank you, thank you.*

Best wishes, Hugh Lovel"

Characters and colors in the book

Colorcodes

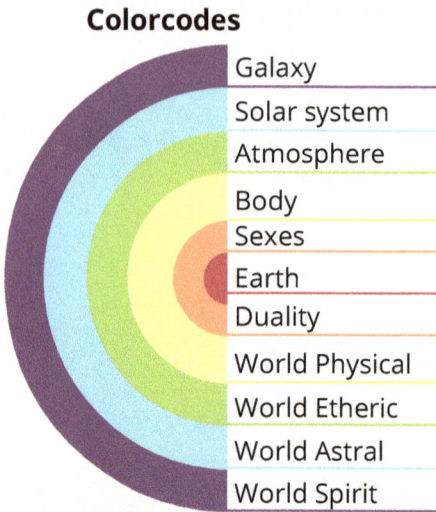

- Galaxy
- Solar system
- Atmosphere
- Body
- Sexes
- Earth
- Duality
- World Physical
- World Etheric
- World Astral
- World Spirit

Explanation

SUL Sulphur
SAL Salt
MER Mercury

Zodiac

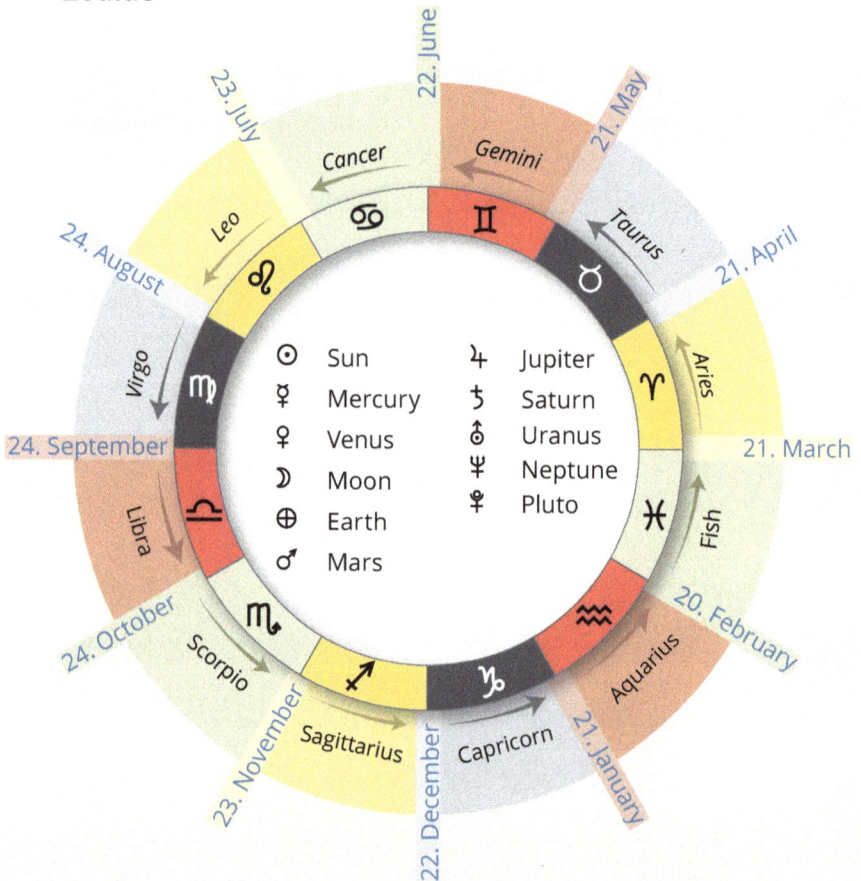

Planets:
- ☉ Sun
- ☿ Mercury
- ♀ Venus
- ☽ Moon
- ⊕ Earth
- ♂ Mars
- ♃ Jupiter
- ♄ Saturn
- ♅ Uranus
- ♆ Neptune
- ♇ Pluto

Zodiac signs and dates:
- Cancer ♋ (23. July)
- Gemini ♊ (22. June)
- Taurus ♉ (21. May)
- Leo ♌ (24. August)
- Aries ♈ (21. April)
- Virgo ♍ (24. September)
- Fish ♓ (21. March)
- Libra ♎ (24. October)
- Aquarius ♒ (20. February)
- Scorpio ♏ (24. November)
- Capricorn ♑ (21. January)
- Sagittarius ♐ (23. November)
- (22. December)

Content

Preface

This is my thought journey of finding the Organisation of Matter. It provides the steps of the story, following on from 'Biodynamics Decoded', through 'Biodynamic Chemistry' and into 'Alchemical Chemistry'.

This story provides the logical progression of universal truths, based upon the principle of As Above, So Below; from the Above universe to the Below of manifest life.

This journey has taken some 50 years, to come to this presentation. I first began to question my surroundings, around the age of 16. Now at my second Saturn return, it is time to collect it all together, as a 'seed', for whatever it may become in the future. I have found the journey continually exhilarating, with the constant flow of insights and discoveries, arising from this 'universal theorem', keeping me at it. I feel blessed with finding it so early in life, as it has held my interest at the level of mild to hot obsession, during this entire period. The practical rewards have made the process worthwhile.

My life has been physically supported by my developments, since 1980, and there is still a broad practical horizon in front of me to explore. While this Saturn return is an ending, where all the parts may now be in place, it is also the beginning of the next phase of my life. I have graduated from my 'university' and its teachers, and have my own experience to move forward with, under the banner of Glenology.

My series of books contain most of the steps I have found necessary to bring enough clarification, for me to feel confident in 'my practice'. I am very glad I have walked this road. It has provided all that it had waved in front of my imagination, all those years ago.
8 May 2013

As the journey through the Atkinson conjecture has developed, it has lead me down roads that can no longer be identified within any of the existing 'isms'. This journey started with Rudolf Steiner (RS), however I have moved to a view that is not compatible, with many of the beliefs his more ardent followers hold. My finally 'getting it' that IT IS ALL ELECTRONIC, has bought some further challenges into that conversation. In practice, my efforts to 'extend' standard Biodynamic understandings and application appear to not be overly welcomed. My efforts with chemistry, have taken my work into a area untrodden by other chemists. I use homeopathic methods for the making of products, however the understanding behind this use, does not conform to 'traditional' homeopathic techniques of diagnosis or disease reference as it comes from Steiner's approach. Yet, the Apop doctors do not use the BD preps as remedies, and have shown very little interest in my chemistry, thus far. I can say I still fit within Astrology, but only because Astrology is the oldest and broadest 'church' there is, with no central authority to define its nature. Simply if you look to the connection of the Above to the Below, you are an astrologer.

The approach presented in my writings now have a character all their own, and can no longer conform to any external reference group for its validity. It can only **be judged according to its own thesis**, and so needs to be called something to define it. Searching for a relevant name is quite difficult as it needs to somehow mirror the work itself without being too impersonal. Thus after some time I have chosen 'Glenology' as its name. This work is a complete expression of my path and efforts in this life, and while it is drawn upon other teachings, the way it appears through my conduit is an expression of my being. It also extends into Glenopathy, given to the strong emphasis on practical application that comes with it. The thrust of my efforts is always towards solving problems, intervening in pathos, or pathologies. So it is with the emphasis upon practical applications, that Glenopathy exists.

It is with a very liberated heart that I offer this book as part of the Glenological series of titles.
24.8.2014

The intention of my efforts are the practical outcomes that arise from them. This book outlines the underpinning thoughts of the logical process I have undertaken to arrive at the diagrams on pages **48** and **152-3**. This provides 'the field' from which practical remedies can be identified. It is a journey, and if taken, it will greatly enhance your view of the ordered world, we exist within. There is a practical application and one that is free and safe for all who wish to use it.

Keep in mind this journey has taken me 40 years, to wander through, and while it will be much quicker for you, I suspect there will be some time needed for your bodies, to digest the many insights and orientations needed for the Periodic Table, to become a living energetic reality in daily life.

This third edition has clarifications, corrections and extensions, most notably of the Lanthanides, more North orientated diagrams, and how 'the Ethers' sit as a reality within the Electronic Being. Enjoy.
3.4.2017

This A4 edition has been produced so the pictures and text can be larger. I find the A4 not so versatile to use, but easier to access in other ways. There are 'refreshments' throughout the text but not huge changes. An important part of the process is the questions. These show the process, the doorways from one step to the next. I have emphasised these in this edition.

Since 2010 I have drawn my pictures using the Earths North Magnetic pole as the reference. Prior to this I used the Northern hemispheres Sun's path as the orientation, as many people do. This means some of the pictures here are reversed to other pictures. When working practically use the Magnetic orientation.

I have also added a section on the different diagnosis methods and uses of the chemical elements, I have experimented with. As part of this preface, is a blog post Hugh Lovel wrote in 2013. As Hugh says I started talking with him on this subject in 2001, and this is what he found. I hope his enthusiasm can encourage more people to take this approach seriously.

I would also like to thank my sons Rimu and Sol for the tasks they took on to make the 3D Periodic Table pictures and sculpture realities. This was an important development to have these forms manifest.
12.3.2022

This English edition is thanks to Ursula who appeared at the right moment to edit and reformat the book, hopefully to make it easier to read.
18.12.2024

My Journey

Initially in my biodynamic journey, I was following RS 'As Above, So Below' suggestion, and was looking at the astronomical reality around me, and applying these forms and ordering patterns to the information in front of me, which lead to 'Biodynamics Decoded'. There was a lot of trial and error in this process, which consisted of running all sorts of information combinations, on diagrams that arose out of joining basic astrology and Steiners lectures. This took some 10 years to work my way through this patterning, which finally arose as the 'Biodynamics Decoded' diagram. There were inspirations along the way, but looking back there was far more effort put into the 'leg work' than the inspirations, mostly due to not really trusting my inspirations fully. Also I was aware of the need to be fully conscious of what and why one is doing what one is doing. Working off inspirations alone, leads to a theory full of faith and very little reason as to why anything is how it is. I wanted a solid base to build upon. I was also entranced by the thought that there is an archetypal patterning in manifestation, and if we followed this pattern, it would provide truthful and useful understandings. I was more interested in proving the cornerstone of this conjecture - that **there are archetypal forms that provide truth** - than simply getting the truths. I believe my overall work has proven this thesis, and especially my chemistry efforts provide the crowning proof, that this is a valid path of knowledge generation, albeit some what slow, looking back now.

When I began working with chemistry, the same sort of 'proving' process was followed. By then I had done most of the work presented in the first section of this book, and had come up with my 'Glossary of the Agriculture' and 'Energetic Bodies Interaction' diagrams. (p. 48). These mostly arose by following the thoughts as they arose, from Bd Decoded. Naturally all along there were artistic steps made, as part of the process.

When it came to chemistry the same process suggested the rectangular periodic table was not an appropriate way to view the elements, as each element, and the planet and solar system etc in which we live, are spherical. So first I worked with a circle, and then later a spherical form for the elements. Looking back, some of my early inspirations eg, where to place the eight arms of the major elements on the circle, were correct in week one, but it took another 10 years of validating that inspiration. The benefit is that today I am convinced the ordering I now present, has great value. So at the 'bottom' of my perceptive process is the 'Astrological model' presented in Biodynamics Decoded' and the organisational rules that arose from this. This is the basis upon which meditation and 'dreaming' can develop.

With regard to 'the dreaming', RS outlines a 3 step cognitive process in 'The Healing Process' (17 july 1924). Behind the rational approach to my investigations there is always those 'breakthrough images' that keep offering the possibility of an answer. Thus there are these two paths showing the way forward. There is the information that arises out of the process of using archetypal patterning of information, which constantly generate many avenues for future research, often by the 'errors' they provide. But also these 'ah ah' moments that come from what RS calls 'condensed thinking'. This is his first stage of cognition, which arises from intensely focusing upon a single thought for a period of time. I seem to do this very naturally. Astrologically this is the activity of the planet Uranus, however a strong Pluto influence adds a degree of obsession to any process.

My astrological indicator of this 'talent' is, a strong Uranus rising, but also Moon in Virgo conjunct Pluto, quintile to Saturn and Mercury conjunct in Scorpio. This allows for 'the bottom line' to be identified very quickly, and a wonderful talent when investigating information. The other significant part of my process is that my Sun is conjunct Neptune and Pallas Athene, while having some tension with a Jupiter Uranus conjunct and Mars. These are a rowdy group who left to themselves have little regard for convention, but have a very imaginative adventurous approach to possibilities and experiences, which lead to understanding. The Sun Neptune in particular is very imaginative and thus I can dream and wander about with images. I believe this has been a facility that has allowed me to follow RS's, more vague and confusing Piscean imagery and very helpful when unwinding his Agriculture lectures, so they could be reformed into something cohesive, as now available in my 'Energetic Activities' book.

It is this area of my being where I can identify with RS second stage of cognition. In the 'Healing Process' RS describes the second stage as, once the 'condensed thinking' of a subject has been achieved, then this image needs to be 'eliminated'. By doing so 'you are descending deeper into your own soul and entering regions that are otherwise only accessible to feeling'. This leads to a state of 'emptiness', which is then 'filled with a spiritual world that surrounds us just as the ordinary physical world surrounds ordinary consciousness'.

My eliminating of the thought, has not been so much a conscious process, as one often borne out of frustration and even exhaustion. Often I feel as if I have come to a dead end, and in a sense 'give up', and release the thought or image. It is usually a short time after this experience that 'the answer' as an insight will arise. Over the years I have become more comfortable with this process and are happier to be 'giving up'.

Recently I have come across the writings of and about Paracelsus. He was the 16th century Swiss 'father of modern medicine', even though he was greatly ridiculed in his time, due to his willingness to follow his own path, and reflect the ineffectualness of his contemporaries, back to them with great vigor. Nevertheless his path has shown the way forward for modern scientific investigation, even though science has neglected to take notice of 90% of his world view. His world view appears to me to be the basis upon which Goethe and Steiner have continued.

His path of knowledge was two fold. Intuition and experience. 'Intuition reveals certain basic facts which then needed to be tested and proven by experience.' Which sums up my process completely. He considered these two pillars could not be separated, they have to work together. Manly P Hall has written a very good short overview of his approach, called 'Paracelsus', and due to copyright restriction I can not quote from it. But it is well worth reading for a more in depth insight into this genius's world view. It is very similar to RS's overall image in many ways.

Essentially Paracelsus, saw all creation as filled with energy, he called light. One form of light we perceive as physical light, and another form of light was more perceptible to the feeling and intuition. This light carried information about the essential nature of things, that was only available to the intuition. He had 'As Above, So Below' well and truly sorted out and commented that 'for every star in the sky there is a flower in the meadow', suggesting he was very aware that the EM radiation of the stars was establishing the 'Spirit' formative force of the many species on Earth. His is a view of energy manifesting as life, in a very similar manner to RS. Thus both men saw the brain as a sense organ that could perceive the 'Cosmic Forces' as they came to Earth. But more than this, they knew that because of the electro magnetic field of the Earth and the Solar System, all knowledge, feeling and experiences that have ever been had by anyone, are radiated into the field around us, and our 'thinking and dreaming' is simply our brains picking up the resonant energies that are already in existence in 'the field'. Paracelsus commented that if all the doctors on the planet, were to die at one time, the knowledge of medicine and healing would not die out. The field would provide the already existing information to anyone who is available to receive it. Thus intuition is little more than us forming a clear thought or image, and then being open to the answer that the 'environment' provides to us. Thus there is no need for formal education as such, but more so, an inquiring mind and a willingness to receive the answers, as they arise.

Here I identify an important part of my process. I have been on the most distant island in the world from the traditional centers of culture and knowledge, and in very isolated provincial regions of those islands. So my 'influences', especially before the internet, have been very minimal, yet when I went to the Mosquita in Cordaba, Spain, the image of the gyro-

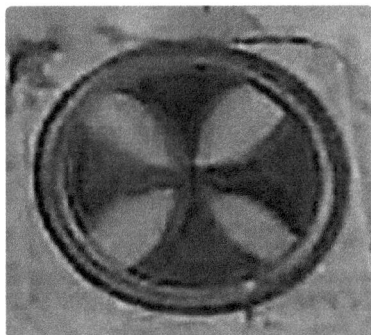

scope, I had drawn from my own investigations, and remains the basis of my diagrams, was on the 10th century floor, a meter below the present floor. Further this pattern is the same as used by all cultures since the Chinese used it 10,000 years ago. Thus I did not need to travel to find the truth, the truth found me, because I asked the questions and listened to the answers.

Manly P Hall also talks of Paracelsus having great regard for all manner of 'beings', that exist in the subtle spheres, along with an acceptance of the role of 'the departed' in effecting and helping the living on Earth.

RS talks of the second stage of his cognitive process, where we have to 'let go', as one also becoming aware of 'spiritual beings' around you, while the third stage one becomes aware of oneself being active amongst these beings.

Here too, I am aware that there has always been 'helpers' on my path, indeed the Sabian Symbol of my Birth Sun position 'A man becoming aware of spiritual forces surrounding and assisting him', which in one version is translated as 'A man in gloom, surrounded by unseen spiritual beings'. The gloom here is not so much my own, but the time in which I live, which by all accounts is indeed a 'dark age'. This brings me back to the Neptune, and Pallas Athene sitting right beside my Sun. I have always been aware that there are helpers in my life. Anyone from the past and even the present can be contacted via the 'intuitive super highway' of electro magnetic energy that we all exist within. The key is a clear intention, a clear question and then the willingness to receive the answer, which is where RS's 'emptiness' comes in. These answers are often challenging, and not easily accepted at first. But as we live with them they make more sense. The intuitive process is available to all humans, indeed we are intuitive beings first and rational beings secondly.

Part of this 'feeling' stage is that one meets ones own inner 'issues'. There is a direct relationship between the 'clarity of the answers' and the amount of emotional baggage one has to get through to hear them. Thus clearing the astral baggage is an important process to pursue, with equal fervor to receiving the answers.

There are also times we are more available for inspiration, than at other times, and a study of your birth chart will identify what your intuitive channel (and astral baggage) is, and when it is getting 'input' from a planetary transit, so your receiving can be clearer and more effective or not. An example of this, is my progressed Astrology, for when the 'Glenological Rosetta Stone' came into view. (see http://garudabd.org/2020/06/03/astrology-of-the-grs/)

So in summary, Paracelsus said it plainly, perception is a balance between intuition and experience. Whatever is received via study and intuition needs to be worked upon practically, so as to provide existential proof of its reality. Indeed so many of the great scientific breakthroughs of the last 300 years followed this path. RS has described this process in greater details and outlined many more of its consequences, however it is essentially the same process. **Clear question, be empty to receive the answer, and go prove it.**

My questions have most always been developed from the archetypal patterning, that arise from the astronomical and astrological 'journey', through RS agricultural and medical indications. This pattern provides the parameters and order, that establishes a lot of what is likely to be real. This does provide a short cut to the good options available to follow, however there is always bits that do not fit. An attitude that the universe knows more than me, has lead to the axiom **'It is ALL right, just where is it right?'** This allows for the bits that do not initially fit, to be used as doorways into other explorations, that more often than not have shown other aspects of the overall whole. This is where 'inspirations' have often proven valuable.

I appreciate my use of the 'archetypal theorem', used by Dr Steiner is an approach not pursued by many other Biodynamic practioners, however the 'conscious' outcomes shown in this book proves its value. This perceptive process is available to everyone. Just be sure to find your inspirations truth in experience, remembering **'a person of faith can believe anything'**.

My Process

Over the years, people have voiced an interest in 'How' I have come to the information I have. What is my Method?

It all starts with the Great Unknown. What is this world we live in, and how can we function within it, in a manner that conforms with its sustainability? We need a way to accumulate information of what is real, understand and order it, then look for possible **'real'** actions that can be taken.

Our subject is Creation, and luckily IT is all around us, and we can all agree it is real. Creation exists. One of the timeless axioms humanity has recognised in its search for understanding reality, is '**As Above, So Below**.' This says if we wish to find an understanding of the order of Life – The Below, then we can look to the Above, for the overall plan. The story of the Above, is Astronomy, and there is plenty of accepted truths that humanity has discovered about the Above. These truths can be applied to what we see in front of us.

This process acts as an **information generator** providing possible pathways forward. Once the knowledge of the Above is applied to the Below, we are Astrologers. Astrology is the study of how the Above influences the Below, in all its various ways.

The more we look, the more we see there is an order to Creation. The common order seen throughout Astronomy is a **Spinning Energetic Gyroscopic Sphere** (a) within which various physical structures are found. In short we exist within an energetic holographic field, filled with objects of the same shape.

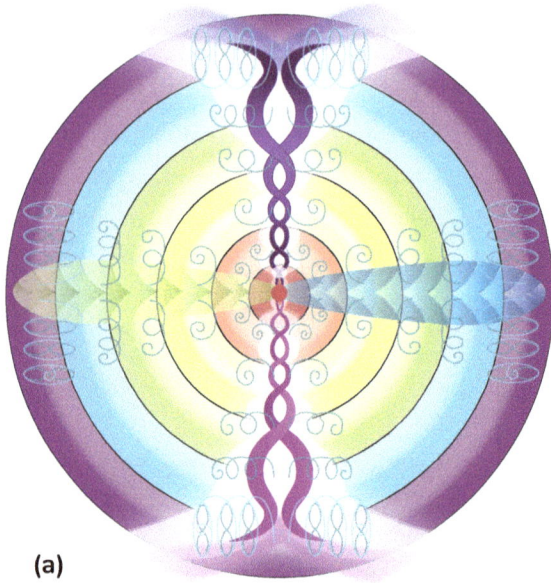

(a)

Once this order is found, it and the fundamental parts of this order can be used to organise any given information about creation, we wish to know more about. As Above, So Below infers the smallest thing mirrors the internal organisation of the largest thing.

Astronomy provides us with a hierarchy of spinning gyroscopes. Not unlike 'Russian Dolls'. While the Universe is huge, it is made up of 'individual' formations of Galaxies, made up of many individual stars, around which planets circle, and in our case, upon which a Atmosphere forms, that can sustain Life. Astrology's long study has stated this (b) as The Zodiac of 12 Stars groups, in front of which, The Solar System of 7 visible Planets move, within which the Earth has its Atmosphere, where the 4 Elements manifest, so that the fold physical bodies can be birthed from the duality, of a male and female a particular species. This IS the backbone of 'The Plan'. This IS the Above. Dr Steiner, and others, provides many functional examples of how the parts of this plan manifests. It is our task to find this plan in all we observe.

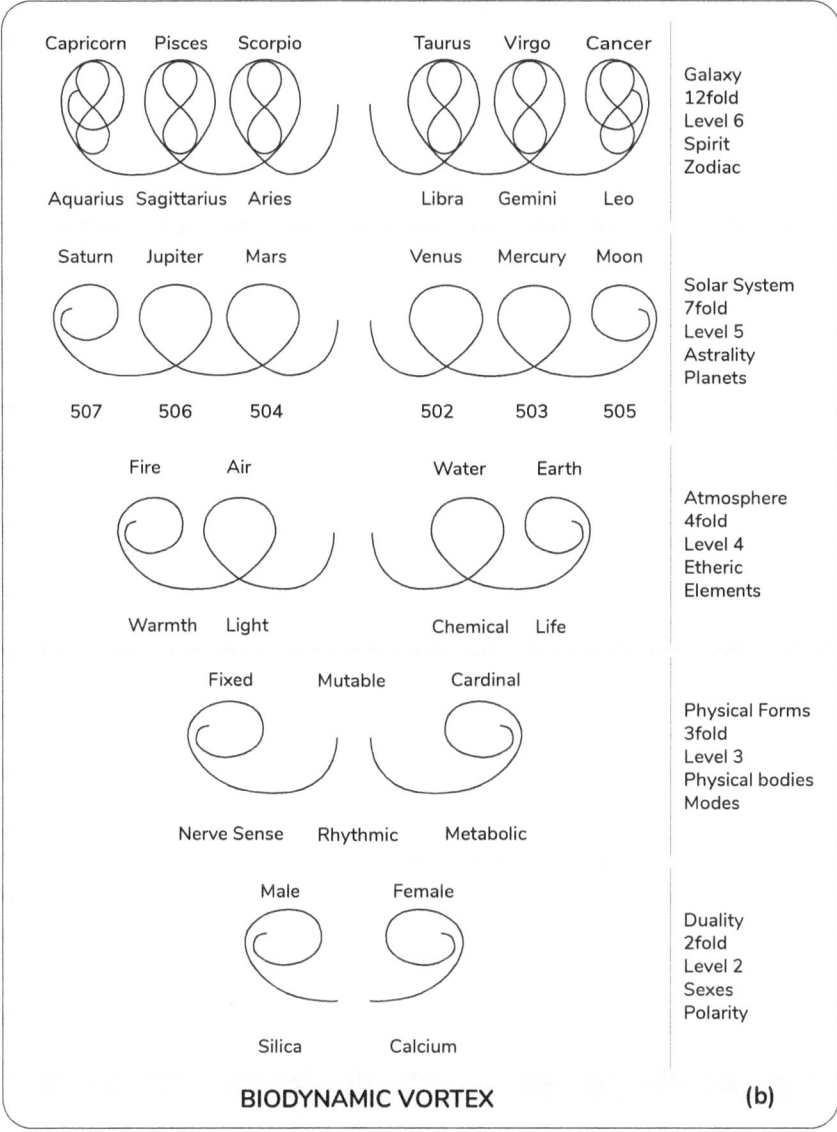

Capricorn	Pisces	Scorpio	Taurus	Virgo	Cancer	Galaxy 12fold Level 6 Spirit Zodiac
Aquarius	Sagittarius	Aries	Libra	Gemini	Leo	
Saturn	Jupiter	Mars	Venus	Mercury	Moon	Solar System 7fold Level 5 Astrality Planets
507	506	504	502	503	505	
	Fire	Air	Water	Earth		Atmosphere 4fold Level 4 Etheric Elements
	Warmth	Light	Chemical	Life		
		Fixed	Mutable	Cardinal		Physical Forms 3fold Level 3 Physical bodies Modes
		Nerve Sense	Rhythmic	Metabolic		
			Male	Female		Duality 2fold Level 2 Sexes Polarity
			Silica	Calcium		

BIODYNAMIC VORTEX   (b)

Through his Goethean Science Dr Steiner has given us a way to generate new information. With his use of 'universal laws' based upon the astronomical reality of the Universe, we can use the observable order we find when looking at the universe, as the 'universal' reference for our observations of Manifestation. The external organisation in Galaxy, Solar Systems and Planets and their Atmospheres, are fundamental organised realities and real energetic dimensions within the Creative Vortex. Physics and the Golden Mean maths that describes this form are the flesh upon these bones. Dr Steiner describes these patterns resonating into the forms of Life, as boldly as in the four Kingdoms of Nature. (21 page 18.) Where each kingdom embodies a further astronomical sphere. The mineral kingdom is just physical Earth. The Plant kingdom embodies the life giving Oxygen filled Atmospheric activities, while the Animal Kingdom adds the sensitive movement we find in the planets of the Solar System , while the Human Kingdom embodies the individualising forces of the Stars.

Dr Steiner used these resonating patterns in many of his natural science lectures. We can see a structure long used by Astrology, and RS talked of as the 12-fold, 7-fold 4-fold 3-fold 2-fold and unity, order. My 'Bernie and Glen' piece adds to this conversation. (22)

My observation method, asks us to collect all the available scientific information on a topic, then look at this information. Look for patterns in the scientific information in front of you, identifying any of these numbers. Once the Number Dimension is established for your cluster of information, we can look for further organisation, within it. In my chemistry investigation I observe three groups of elements, each with a different number. There are 8 groups of Major elements, 10 groups of Transition elements and14 groups of Lanthanide elements. This suggests the numbers 4, 5 and 7 can be organising principles for these elements.7 suggests the 7 planets provide the backbone and that there may be a inner polarity between the one group of 7 elements and the other group of 7 elements, within the 14 Lanthanides and so on. I then used melting points, valances, atomic weight and the elements uses, to further define an order that can be further related any other information I already have related to the number 7, within a circle. (1) In this way we can build up reference points for establishing possible truths that can be further researched. We can ask how does the inner order enliven the information and what insights have been generated through that exploration? Trials of various kinds have shown the Lanthanide elements act as this process predicted.

This process has led to a large Astrological Equation, that moves according to the context it finds itself in. The context always has some relationship to an Astronomical basis, so has a basis connected to truth. However the Universe is a big place and there are many context changes. Keeping an eye on the astronomical reality is the shining star in the night. The Astrological rules built upon Astronomical truths, provide some boundaries to form our findings into things worth investigating. Astrological observations are based upon the Physics of Creation.

This new information should be considered speculative, however the speculation is based on a proven theorem. So it has a very high potential of being right, and is thus worth pursuing further. We must look for ways to prove this speculation, or not. Keep doing this.

I see this activity as Dr Steiner's first stage of Perception. He then describes a second stage as what happens when our intense study and observations reach a peak, which in my case is usually accompanied by some frustration. At this point I release or give up on the thought form, and into this space flows new possible ways of approaching the subject. An answer to the question that was unable to answered before, often appears.

We can imagine this process as our Internal Spirit, through focus thought, enlarging in size and pushing out into the surrounding environment. Our 'environment' is created by the real energy coming from ALL the billions of Stars. This energetic holographic field is called the Cosmic Spirit sphere. It is within this energetic field, that the World activities of the planetary Astral and atmospheric Etheric influences, can do their further organisation. When we actively think and imagine thought forms we are pushing out into the World Spirit sphere of our Solar System, at least. We are consciously looking for inspiration for our questions, pushing in the Collective Unconscious, held within the Sun's field, with our meditations on our subject. Most people can receive some direct help from this activity. However when we release our focus and the push outwards, the World Spirit and Collective Unconscious push back and flood into the empty space in our consciousness, left by our thought form. The answer is supplied out of the Universal reasonance responding to our thought development. Thought is indeed a formative force and the universe responds. Everything is in movement and in response to everything else.

The third stage of RS perceptive process is when we can meet the energetic individuals in the surrounding field, who are focusing these Cosmic Thoughts back to us. Here personal conversations with these energetic individuals can be had. These Beings can be of all manner of nature, from dead people through to Elemental, Planetary and Star entities. Whoever you wish to communicate with is usually available. Just ask. Receiving what they have to say is often my difficulty.

This is the process of 'knowing all things'. Every thought and feeling ever had by anything on Earth, leaves an impression within the energetic field of the Earth. If we can become conscious of how we are embedded within this field, we can access Inspirations, Imaginations and Intuitions. Ask and the field replies.

When this process become fully conscious one is 'Enlightened', and is one step beyond Pluto and at the sphere of Persephone, which Dr Steiner calls 'Solid Land' in lecture 3 of Occult Science. Creative visualisation or directed intentions leading to manifestation, becomes almost instantaneous, in this sphere.

During all these stages we are generating really good new information. Dr Steiner suggested that his followers join together and listen to each others insights as 'spaciously' as possible. His path is about discovering new truths which work with the Universe, to produce more ordered and sustainable Life. This is a new frontier and we have the tools to explore it consciously and safer.

Hugh Lovel's experience

Fruit Tree Problems and Solutions
Hugh Lovel (after the 2013 Advanced Quantum Ag conference)

Dear K and H, It all seems mysterious at first. I gather the two of you picked up on Glen Atkinson's masterful grasp of the forces of the surrounding universe and how the various types influence our lives and environment. I've been knowing this bloke (Glen), who originally is from the Sunshine Coast of Queensland, for close to 20 years, and I think we were enormously lucky he could attend and participate in our Advanced Course. While I cognited more than 40 years ago in quantum chemistry class that astrology is valid as a means of identifying how the universe, both large and small, contributes to our lives, I am nowhere near the astrologer that Glen is. Listen to your Dennis Klocek Macro/Micro lectures on your course materials disc. Dennis is another such astrologer/biodynamic scholar like Glen is. Ten years or so ago Dennis used to do a weather forecasting website where he used astrology to predict hurricane paths and intensities. Living in the southeastern US I sometimes used to watch his predictions, which were far more accurate, detailed and prescient than the US weather bureau's.

Mysterious as it may seem at first, there is a science to all of this, and we can sort it out using the tools we have. Considering that Glen is in the business of selling agricultural products whose active ingredient is water, it is good that he also tries to share the understanding of dynamic geometry (astrology), the periodic table and biodynamics that he used to develop his product line. Besides, the important thing is the understanding, and I truly do respect all the work he's done.But in dealing with the broader picture, let's sort out our problems and how to deal with them at every level and in every way possible using the tools we have, which includes Glen's insights into how the preparations work with the different planets, constellations, elements, ethers and dimensions.

Glen and I did a workshop together in the Willamette Valley in Oregon back in 2001 or so and at the time we had a huge private discussion about where the various elements of the periodic table fit into what he calls his Gyroscopic Agriculture model and what he calls The Apple Of Life. At this point I realized that what previously had been a mystery to me regarding the geometry of the electron orbitals that a century of nuclear physics (since Bohr) mapped out could be resolved through mathematics of dimensions (geometry) if hydrogen and helium were treated as one dimensional vortices and each successive octave (based on lemniscular motion) was seen as adding a dimension to the picture.

Don't worry about the maths, but this is what gives the various elements of the periodic table their functions in regard to each other. It means that carbon, a two dimensional element of surfaces (a chemical element), has the information of all the possible forms in the universe writ on its surfaces, or if you prefer, etched upon the planes of its particle vortices and their potential interactions.Now carbon is only two dimensional (chemical) while silicon is three dimensional—which makes it a physical element. Calcium is four dimensional, which makes it etheric. Following calcium are the 4D transition metals that are key for all the enzyme/hormone processes associated with living organisms—all forth dimensional. How can we use this sort of thing? We've had a big problem in agriculture with identifying the causes of things. The result has been a long series of patches that only treated symptoms and left the underlying causes untouched. With the above sort of information we can identify causes. It shouldn't be any wonder that good old 3D silicon (along with sulphur, also a 3D physical element) forms all the cell walls and connective tissues (along with hydrogen (spirit) and the chemical elements) that provide physical structure for living organisms.

If we want this structure to be strong, resilient, durable, robust, etc. we have to ensure the availability of silicon (in partnership with magnesium and phosphorous). The chemical elements of carbon, nitrogen, oxygen support this with boron and fluorine as chemical co-factors in ensuring the fluidity and functionality of Si, Na, Mg, P, S and Cl. So the chemical elements (carbon and its mates) lie behind the creation of the physical structures composed of silicon and its mates. (I hope I haven't lost you yet] The point is the internal, life (etheric) processes that allow for growth and reproduction of these silica structures is the business of calcium and potassium and their co-factors (lime and the transition metals).So when we have a physical, structural problem, such as the skin of a nectarine or cherry or the stems that hold them on the tree it has to be silicon. We don't have to think twice about it. Structural problems are silica problems. On the other hand, when we have an internal problem such as flavour, nutrition, growth, sizing in fruit development, anything metabolic, etc. it is a calcium problem. It has to be since these 4th dimensional processes go beyond the physical structure. But since what takes place on the insides of cells depends on what hold them together and feeds them from the outside, the lack of sizing in the early development of fruits depends not only on the nutrient supply of calcium, carbohydrates and amino acids, but the containment and delivery system that goes back to silicon and its mates. So a lack of boron and/or silicon will result in poor sizing in the early development of fruits (calcium) and poor filling out with sugars and flavours (potassium) in later development.

Why do we have fungi, on the one hand, or insect problems on the other? One is a deranged etheric situation and related to an excess of watery lunar forces; the other is a deranged astral situation and related to the dry/warm solar forces. Whether these problems occur above ground or below ground also tells us things that relate to the seasons and conditions where these problems originated, which may greatly precede the time period when they show up. But we've got the tools to sort these situations out and remedy them at the cause rather than trying to patch them through once the problems show up. This, for example, should show Kym why splitting occurs in cherries. It isn't all that much a problem of having a rainy harvest. The cherries split when their connective tissues and membranes are weak. This is a problem of stress and lack of silicon/boron/fluoride in the early, early structural development of the connective tissues of the fruit. So forget fixing it once you see it. That's not where it occurred. It occurred where there was a disconnect between the warmth/light/silica forces associated with Saturn and the chemical/life/calcium forces associated with the Moon–combine too much watery nitrates in the soil along with cloudy, cool days in the early spring following fruit set and you have a problem.How to address it? In this case it is VERY important on the one hand to spray horn silica on the soil in the winter to build up warmth and light and silica in the soil.

On the other hand it is VERY important to have a good annual cereal/legume cover in the orchard and that the cereals suck up nitrates and deliver amino acids and lime as they are digested. Judgment needs to be used concerning mowing of the cereal cover (and probably planting a summer cover at the same time) following fruit set so the weather and nutrient flows are managed for optimum nourishment of the juvenile fruit crop. Also keep some phytolacca and nettle ferments going along with appropriate herbal BD preps to use in fine-tuning this dance with the weather. Enough for now. Do you guys see where I'm going with this? You can't even think about these things – they would remain a mystery – without having a framework such as astrology to make sense out of the warmth and light activities/elements of saturn and how this works with the chemistry and life activities/elements of the moon. For certain I don't know much about specific problems because there are so many and they vary considerably from place to place and year to year. **All I know is there are ways of sorting out the causes and shifting the situations at their points of cause.** Sometimes it takes me a while to sort things out, and a discussion would be very helpful. So when you have problems, how about if we have a go at sorting them out as a group discussion? Then we'd all learn. And when you are at the top of your game you'll keep things balanced and never even get new problems that sweep the industry. As you can see, I have a lot of time for you guys.

Best wishes, Hugh

The Overview

In his lectures on natural science, Dr Steiner encourages us **to look to the macrocosm, and then find the similar patterns within the microcosms,** we have around us as life forms, if we wished to see reality.

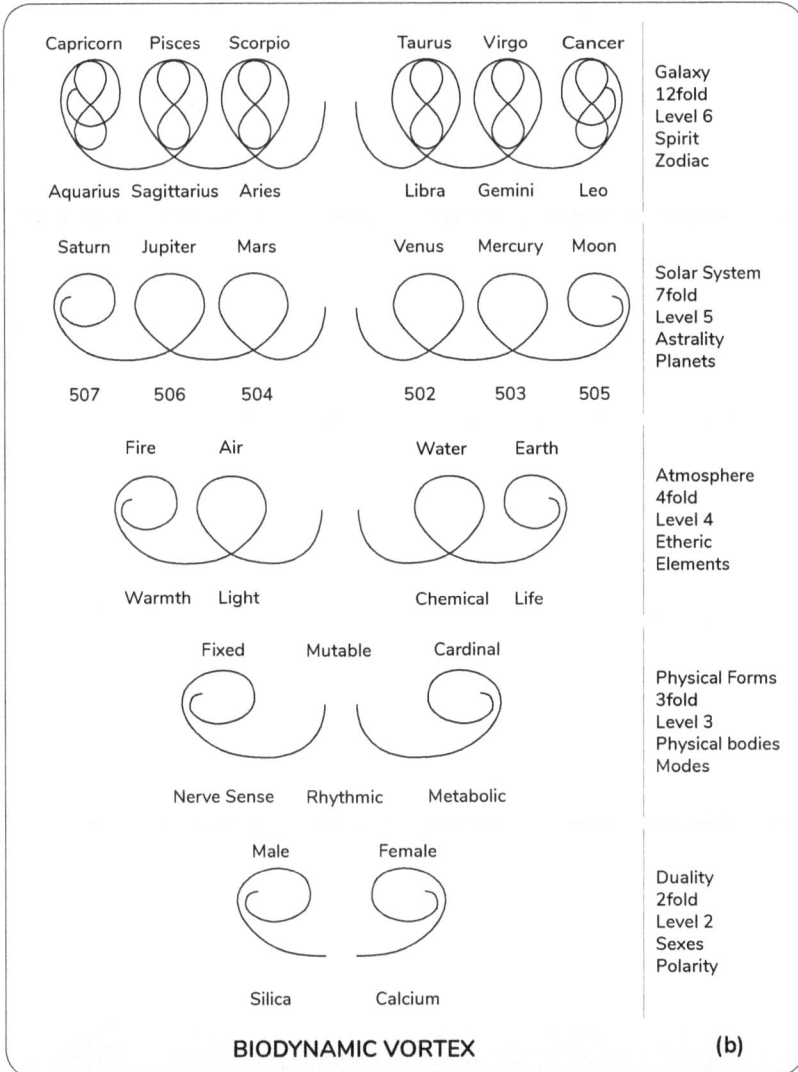

Capricorn	Pisces	Scorpio	Taurus	Virgo	Cancer	Galaxy 12fold Level 6 Spirit Zodiac
Aquarius	Sagittarius	Aries	Libra	Gemini	Leo	
Saturn	Jupiter	Mars	Venus	Mercury	Moon	Solar System 7fold Level 5 Astrality Planets
507	506	504	502	503	505	

Fire   Air   Water   Earth

Atmosphere 4fold Level 4 Etheric Elements

Warmth   Light   Chemical   Life

Fixed   Mutable   Cardinal

Physical Forms 3fold Level 3 Physical bodies Modes

Nerve Sense   Rhythmic   Metabolic

Male   Female

Duality 2fold Level 2 Sexes Polarity

Silica   Calcium

BIODYNAMIC VORTEX   **(b)**

In 'Biodynamics Decoded', I showed how the basic forms apparent in the formative processes of the Galaxy, can be used to understand the formative patterns of the activities outlined in his 'Agriculture' lectures. One of the fundamental patterns of life, **the vortex**, was used to present an overall diagram of the biodynamic information, as 'Biodynamics Decoded'. A detailed colour version of this simple diagram can be obtained for free at www.garudabd.org. I also call this diagram the Biodynamic Vortex.

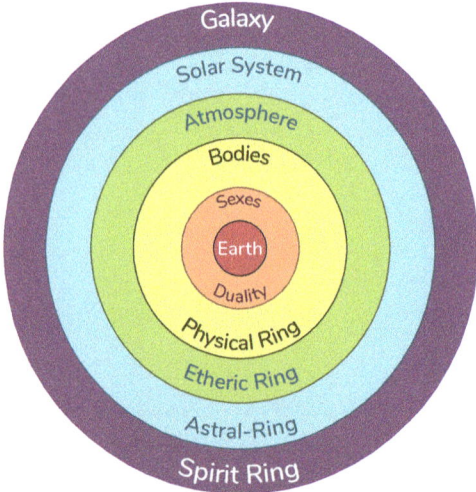

The Vortex however is only one part of the overall galactic form, which is really a spherical gyroscope. More commonly called a Torusian Individuality these days. The Vortex is found, most obviously, on the vertical axis of the gyroscopic form. So while it was useful to present the biodynamic world view as a single vortex, it is more correct to **expand this picture into the gyroscopic form**.

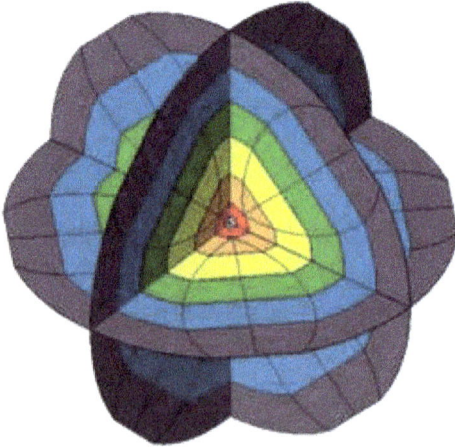

The essential element of the vortex are the layers. Astrology and Dr Steiner outline these as the 2 fold, 3,4,7,& 12 fold layers.

In the accompanying diagram the 12 fold – Galaxy (Mauve), 7 fold is the Planets – Solar System (Blue), 4 fold – are the elements of the Atmosphere (Green), 3 fold is the structure of physical bodies (Yellow), and the 2 fold is the primarily male and female nature of biological organisms (Orange), one fold is the Earth itself (Red), are shown. The bottom picture expands this from a circle to a sphere. These layers represent the real but different dimensional layers of our environment. BD Decoded shows most of the references you are likely to find in the Biodynamic literature, placed upon this vortex. By following the rules of the vortex, many relationships and activities between the parts are revealed.

The next step though is to take these conglomerations of information and reorder them into a sphere. The first thing that can be done is to identify that within the spherical gyroscope, we can imagine that from the Earth in the centre, the layers are experienced as being like the rings of an onion. Each layer is its own sphere within the other spheres. This is the astronomical 'cosmic' reality within which we exist, and is **'what is there'**. This is all very well, however more organisation can be bought to this image.

Matter's Electronic Organisation

Once movement in the Cosmic static electron field occurs, polarised Electrons begins to spin and a magnetic field develops, which in turn generates an electrical charge, within a spherical field. This is called the **Torusian Electronic Being**. This field is the dominant organising principle of all matter present within its field. Given everything in creation conforms to this form it is to our advantage to get to know it well. I use the formula Cr = Mv + T (Creation = Movement + Time) to indicate this process. (I appreciate my use of the word Electronic is not the normal usage, however I do not have a word that contains all the forces active within a spinning gyroscope, some of which, along the DiMagnetic scale, do not support electro magnetic activity.)

Not being much of a physicist, I can not give you the maths of this phenomena, however the people I think can are Nikola Tesla, Walter Russell, Dan Winter, and Nassim Haramein.

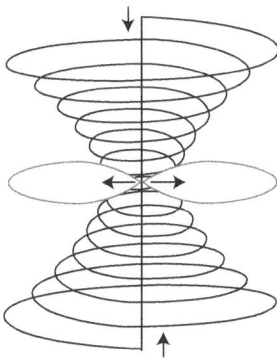

My approach is more observational of the phenomena available to me. So... **Due to the movement of the object**, the resultant magnetic field produces positive and negative poles. These poles are characterised by the formation of vortexes, which can be seen in the magnetic field of Earth at the North and South poles. These poles, with their spinning vortices cause a flow of matter, moving towards the center, and a flow of force moving from the center outwards to the periphery.

The matter, moving towards the center, is consolidated, or consumed in the center, (i.e. is it is a planet or a Sun), before being spun off again along the horizontal axis of the gyroscope. This horizontal plane of activity becomes manifest, and seen as the substance of the Milky Way galaxy, or the planetary sphere of our Solar system. There is evidence that when something reaches the edge of the sphere it is drawn towards the pole again and recycled through the middle again. At the galactic level I understand it is accepted that the star Archturus has done this journey. The Kieper belt and Oort field of 'rocks' beyond Pluto, also suggest that debris is 'excreted' out of the solar system, via the horizontal plane.

We must never forget that **everything is moving** very quickly. For example on the Galactic level, our accumulative speed on the Earth, has us moving at 118,000 mph through space (Earth rotation + Moving around the Sun + movement around the Galaxy). We generate serious magnetic fields along the way.

The horizontal plane has a interesting feature that can be most easily observed once a star has gone supernova. "Going supernova' is when a star reaches such an intensity that it explodes all of its inner contents outwards, thus filling its 'atmosphere' with cosmic dust. This gives us a look at what the shape of this 'atmosphere is.

In these two examples, there is an obvious vortex form, but also what appears to be a lemniscate form, as if there is a pulsing between the two halves. This cosmic image has a earthly human correspondence within the human Rhythmic system, in the relationship between the heart and lung system, which pulsates at a 4:1 ratio across the middle 'horizontal' plane of the human.

So added to the double vortex image of the vertical plane, the creative form has a horizontal plane. Which with its vortexing lemniscate, makes the form of the **Greek Cross**, common in all the world religions. This form has been central to humanity for over 10,000 years, and this cross form is the basis for the dominant physical structures of the gyroscopic spherical form, within which we live.

When looking at this picture, not only do we have the nature of the vertical and horizontal planes, but we must also surmise about the qualities of **the spaces in between** these four basic axis. By doing so we meet the double cross.

The Double Cross

It does not take much observation of cultural imagery, before one comes across a wide array of images, based upon the 'double cross'. It can be found in ALL major cultures from the Chinese, Vendanta Asians, the American Indians, The Maya, through the Persian / Islamic culture and throughout the Christian culture, even St Peters Cathedral and Square, in Rome are built upon it. The Chinese and the America Indians say they have used this as a central form for 10,000 years. The question becomes, **what is this image**, and what can be the basis of its significance.

In most cultures there are two main references of orientation. The first is based upon magnetic North and so we have North, South, East and West, and on a compass (at the equator) we have North pointing along the horizon line. Thus south is towards the opposite horizon point, while East and West are on a 90 degree axis to this, but this is not quite true for all latitudes...

The next reference system we find is related to the seasons. Many ancient monuments are aligned to the Solstices and Equinoxes, as seen from their hemisphere. The Equinoxes are marked on the horizon for where the Sun rises on the 21st March and 21st September, while the Solstices are where the Sun rises, at the horizon, at mid summer and mid winter. A good example of this can be seen in St Peters Square in Rome (12, p. 29). **This orientation is thus focused on the Sun, and not the magnetic orientation of the Earth.**

Both of these reference systems however appear (in our daily life) to be related to the horizon line, and therefore it is very easy to see them as similar, until we look at them within the spherical reality in which they actually exist.

In the EM gyroscopic form (p. 31) the north south axis is based upon the primary poles of the 'magnet', with their vortexes, at the top and the bottom. The vortexes on the Earth, allow for the Aurora to form, when the Solar wind passes by them.

From here, we then have East and West, at a 90 degree angle, based upon where the Sun rises and sets. This provides two axis of the sphere. The 'Major' axis (N/S) and then the horizontal plane marked by the position of East and

West, at the horizon. There is however another axis needed to make a sphere. This is the second vertical axis which also crosses through north and the horizontal plane, at a different 90 degree angle to the 'Major' plane.

This vertical axis may not have had much relevance, if it was not for the fact that all the planets, (and thus the Sun) are all lined out along the horizontal plane. Depending on where you live on the planet, the horizontal plane makes an arc across the sky, at varying distances from the horizon. Where the Sun reaches its highest point in the sky, is where the second vertical axis crosses the horizontal plane, and this point is called the Zenith. The point opposite is called the Nadir.

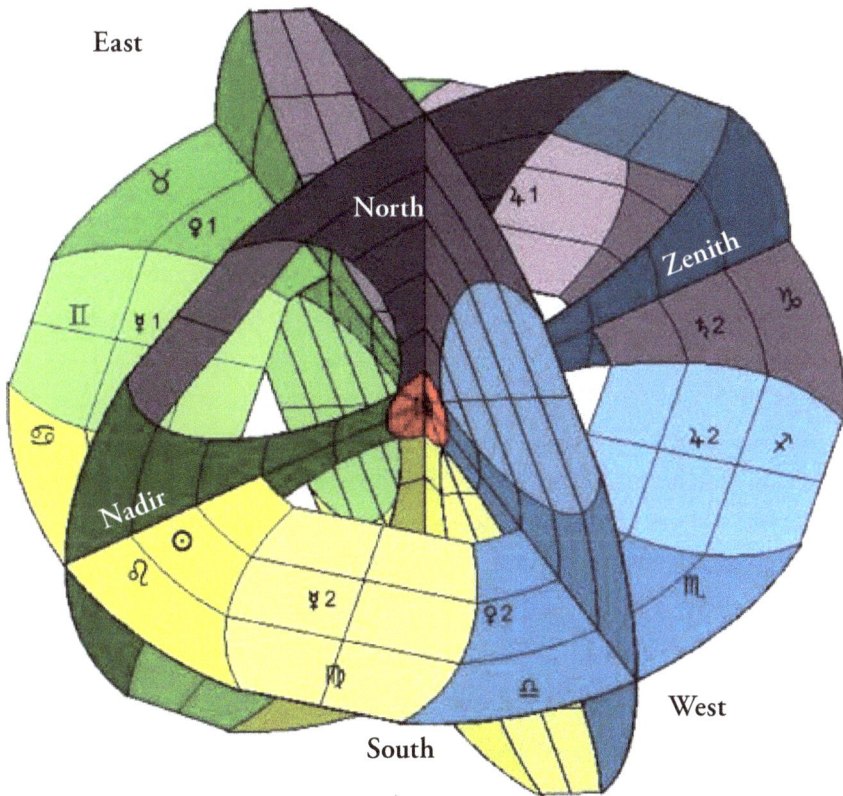

This diagram is as if you lived very close to the north pole, in the northern hemisphere so the Magnetic north is directly above you and the Sun, moving along the horizontal plane, passes just above the horizon, when looking 'South'. As the horizontal plane spins and the Sun reaches its highest point of travel each day, it reaches its Zenith, before tracking lower towards the horizon again. As you travel across the planet towards the south, the north pole will move to the left and the Zenith point will rise up in the sky, until, when at the equator the north point is on the horizon and the zenith is directly above you. Continue moving southwards and the north pole goes below the horizon and the Sun's path again begins to lower, but now is seen to be going across the 'north' part of the sky. So the Zenith axis points are very relative to where you are on the planet, and also at what time of year you are viewing them. Especially when one is at mid to high latitudes, the angles between the Zenith axis and the East West axis change throughout the year, due to the Earth being on a 23 degree tilt in relationship to the Sun's ecliptic.

Because of our 'cultural' orientation to the Sun, from a northern hemisphere perspective, the horizontal plane and the Zenith axis have gained a dominant role in our awareness, over the North South axis. Even though in the southern hemisphere we tend to see them as the same thing.

The main differentiation I wish to highlight here is that the **N,S,E,W axis are a completely different 'reality' to the cross activities of the horizontal plane**. N,S,E,W are always in the same relationship to one another, while the Zenith Nadir East West is a subjective experience based upon where you are on the planet, but more importantly it is a phenomena related primarily to the horizontal plane. Thus we have a **objective reference and a subjective reference**, which has lead me to call the primary 'magnetic' cross the **'Cosmic Cross'** and the secondary 'zenith' cross the **'Earthly Cross'**. This duality between Cosmic and Earthly, Objective and Subjective, External and Internal is a theme that keeps arising, and one I explore further in "Biodynamic Questions, Astrological Answers".

The 'Double Cross' diagrams (p. 30) are all two dimensional representations of these three dimensional activities. **The Cosmic Cross is what is real.** It is a manifestation of real spinning bodies, while the Secondary **Earthly Cross areas will respond** to whatever movements these primary cross arms make, in the same manner that Earthly things respond to the broader cosmic activity. Therefore within the 2D double cross, we can expect to see the interplay between the

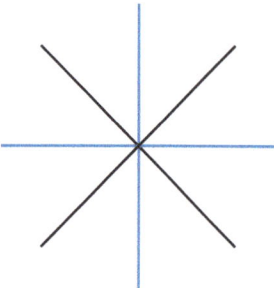

Cosmic and Earthly activities, or **the external World and internalised activities**.

Discussions I have had with American Indians suggest their medicine wheel, acknowledges the nature of the four 'winds' of the primary axis, as well as giving images of the activities for the in between spaces. They are considered very restless spaces and not places to spend a lot of time in. They orientate their wheels off the Sun. So North is Cold and Earthy. In diagram 1A (p. 30), they are white petal shapes.

Macrocosm (External) and Microcosm (Internal)

Mineral Kingdom
Internalised Physical Body

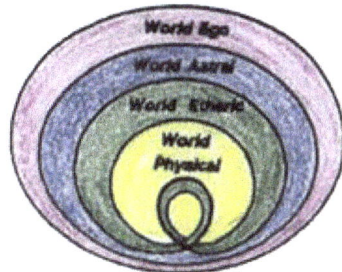

Plant Kingdom
Internalised Physical & Etheric Bodies

Animal Kingdom
Internalised Physical, Etheric & Astral Bodies

Human Kingdom
Internalised Physical, Etheric, Astral & Spirit Bodies

Dr Steiner (RS) and Paracelsus before him, suggested we always make reference back to the macrocosmic, and look for the harmonic resonance that connects the large with the minute. They emphasised that life is a manifestation, achieved through the internalising of the activities of the external spheres, and that the various kingdoms of nature occur due to the differing degrees of this internalisation process.

RS had to give names to the electro magnetically energetic activities found in these four great astronomical spheres, and he chose to go with the names that had been given to them since Hindu times. He called the Galactic activities , Spirit, as these are the constant and formative impulses being beamed at us from all directions. For the Solar System energies he called these Astral, the Earths atmospheric energies are called Etheric and the Earth itself is called Physical. He commented in the 1923 medical lectures that 'we have to call them something'.

The mineral kingdom is lifeless matter, while the plant kingdom internalise the energetic nature of the atmosphere of the Earth, called Etheric. The animal kingdom adds the solar system energies, to develop proteineous flesh and organs. While the Human kingdom internalise the energetic activities of the stars of the Galaxy, which produces individuality and rational thought found in humans. Life is indeed then a microcosmic image of our macrocosmic environment. These diagrams are a visual representation of this enfolding process.

Many of RS followers believe these astronomical energies are something more than just electro magnetism (EM). While there is a EM component, there is also an accompanying "Etheric' activity that stimulates Life processes via the Di Magnetic activity. On page 88, I discuss how the non EM, Di Electric / Scalar / Etheric sits within the gyroscope. **There is no escaping the EM reality of Creation.**

For me everything in creation is Electronic, and a result of the movement and order associated with the electronic beings, we have in creation. There is nothing spooky about star forces coming via the planetary and Earths EM fields to reach the physical Earth, which in turn then radiates energy back outwards into the cosmos. This is all accepted fact. We do not need religious belief for this world view to make perfect sense. **'A person of faith can believe anything'.** Once the faith barrier is crossed there is no end to the extremes of belief that one may entertain.

So when dealing with reality, we are challenged to find how the **outward radiation from the Earth manifests in life forms**, while interacting directly with what is coming inwards from the Cosmos. We can then take these considerations as the basis for growing plants and husbanding animals.

Firstly, it helps to become very aware of your connectedness to the external cosmic activities. While this might seem a difficult task, it can be easily answered with an open minded experience of having ones birth chart, (the planetary positions at your birth), reflected back to you, by an experienced astrologer. In this experience you can see in your own personality, the connections to the planets, and hence your astral body. One then asks the question, 'How can someone who previously knew nothing of you, be able to describe in such detail the very private wanderings of your inner life?'

Further to this, explore the events of your life, and how these were reflected in the planets movements at the time, followed by predictions made of your future, with specific dates of potential specific outcomes. In this experience you are consciously mapping the connections of the planetary sphere to your internalised astral bodies movements. **Here you can experience how you are intimately connected with the macrocosm**.

The etheric body connection to what comes from above, is harder to make conscious, without a study of the Earth EM field fluctuations marked against human health phenomena. I have not done this, however the correlation of the Schumann resonate of our atmosphere – 7.8Hz – to our energetic resonance shows our 'union' with the atmosphere.

Experiencing your internal etheric body is as simple as watching the difference between your experience of being tired, and how you feel when you wake up. Sleeping is the time we recharge our etheric body. We should feel 'fresh'.

The Spirit is experienced as 'the controller'. That bit of you that when properly incarnated gives you a feeling of being in control of your life, with the ability to clearly define the options in front of you and to make a calm choice. When it is not incarnated properly, things can seem chaotic and you can feel pushed around by life.

All these activities have their manifestations in soil, plants and animals and Biodynamics is the study and practice of this.

FIRE
Nerve Sense

Ego   **H** G gene

Water   **O** (Etheric ←X→ Astral) **N** Air

A gene   **Circulation**   **Respiration** C gene

T gene   **Physical C**

Earth   **Metabolic**

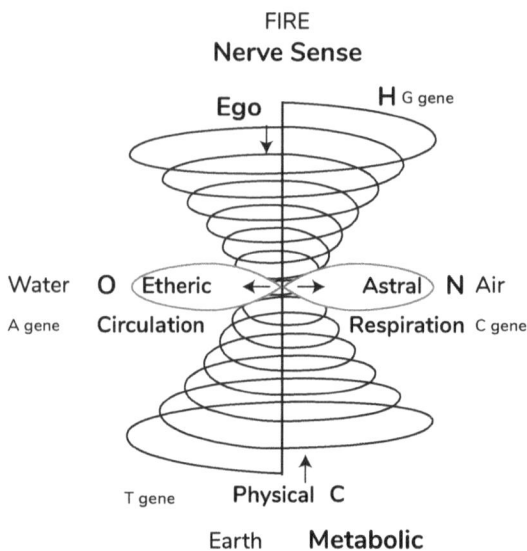

The practical challenge then is to **find these relationships between the inner nature of the plant or animal and its external environment**. Biodynamic scholars have made extensive studies of these correlations (see bibliography).

Manifestation is what happens in-between the Cosmic and Earthly activities. When growing plants the external reality rules. The condition of the soil, moisture, light and warmth are THE dominant factors influencing plant growth. Without these there is nothing. The internalised activity of the plant, is governed by the status of the external environment, it finds itself in. It is tossed between the varying activities of the primary external cross and its own internal processes, that did originate from these external sources.

Dr Steiner gives very good details of how these 'cross axis' work together **in** living systems, and most clearly when he talks of the Human systems, in his medical lectures, especially those in October 1922 (17).

The way the activities work externally and internally though are somewhat different (see page 54). In the **bodies external relationships** outlined above, **for the plant**, we have the Earthly element of Physical / Earth, and Etheric / Water working from below upwards, polarising with the Astrality and Spirit working downwards. This can be seen practically in the Earth and Water elements residing below while the Light and Warmth elements come from above.

For the **internal working of the bodies**, we see greater polarisation taking place. Life arises out of movement and where there is movement we find first the spiral and then the gyroscope. In the gyroscope we find **polarisiation of opposites**, such as the north and south poles of the EM field. Within living beings, the vertical axis is maintained through a dynamic polar relationship between the Spirit and Physical Earth activities - 2 fold. Once these reach a certain intensity of interaction, the center forms, creating a 3 fold form. The center then divides into a polar relationship between the Etheric and Astral activities (4 fold) on the horizontal plane, and the gyroscopic lifeforms become self sustaining. As with my use of the 'Biodynamic Vortex' picture, this gyroscopic picture is archetypal, being based upon galactic formation process. This picture has both the external and internal two cross patterns, placed on just one diagram. It provides an easy way of showing the relationships between the various parts of life.

This basic cross can be enlarged by placing all the references RS gave for the various physical processes (see p. 54). This basic gyroscopic pattern shows the elements of the atmosphere, Fire and Earth etc, the spiritual bodies, Ego / Spirit physical etc, The elements of protein chemistry, DNA and the physical systems of a physical body, all according to the energetic activity it is a manifestation of.

Seasonal Cross and the Internal Arms

The next step in the development of this diagram is how to exactly allocate the internal arms. We already have it that the internal manifestations will be in the 'in between spaces' of the primary cross, but what body should be in what space. It would appear reasonable that the internal arms will be directly related to the external arms, with the astral light and spirit warmth being above the horizon, while the Physical Earth and Etheric Water are below. The next question was which way to spin the arms, for the 'right' relationships. A twist, one step anti-clockwise, makes for an obvious symmetry. This makes the internalised bodies – Spirit and Astral working from above the horizontal axis and the Etheric and Physical bodies from below the horizontal axis, which is an image of their external archetypal positions in nature.

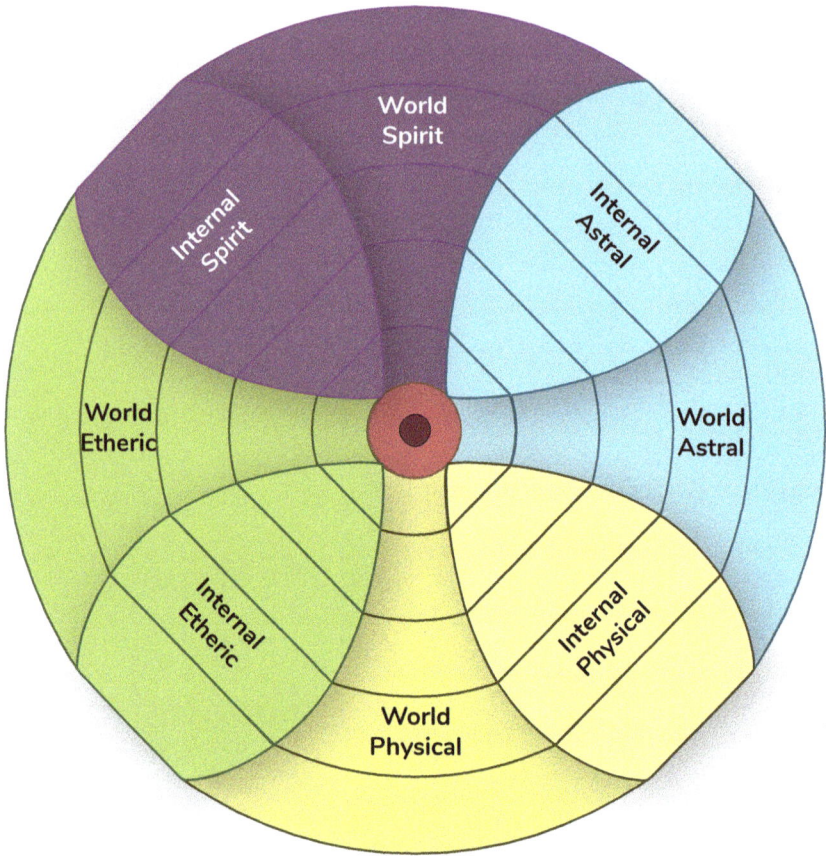

The Energetic Bodies in Nature

There are two main organisations of the way the energetic activities present themselves. The first – **the Rings picture**, is how the energies sit in the cosmos, and so I call these 'Cosmic' activities – strictly speaking the **'Cosmic activities'** also includes all the activities beyond our Galaxy. The second organisation is how the energetic activities organise themselves, when they move, and become active in living processes. This is the **gyroscopic diagram** with the **World and Internal** activities on the previous page.

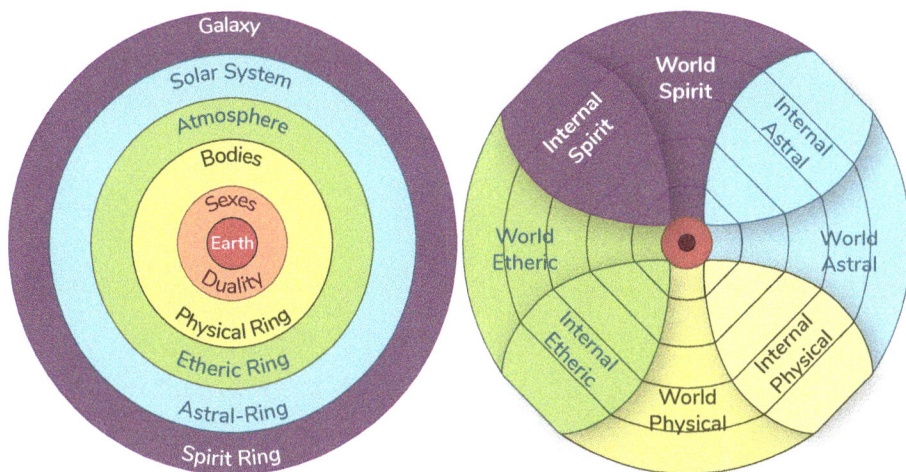

It is very difficult to find the clear differentiation of these two stories in amongst RS words. One of the problems with the Agriculture course is, he was telling these two stories at once, and tended to interweave them together, through the whole of the course. This interweaving has caused endless problems for future generations, so that today it is generally accepted that this course is incomprehensible. Therefore large parts of it - the main theological tracts – have been 'officially' set aside. In their place, stories of the Ethers and Elementals, from the 'Man as Symphony' lectures, and other often Christian based theologies, have been expounded. Given the modern Biodynamic worldview is not properly based upon the Agriculture Course, it has no common theology throughout the world, as individuals are free to makeup whatever story they wish. It is often seen as a chaotic system of quasi religious beliefs, with little relevance to a modern scientific humanity. As one travels about the planet this basic anarchy has one group saying one thing and another saying the exact opposite. Given the 'social state' of the dominant organisation, where the religious Anthroposophical stream holds sway, I can not see any change in this circumstance, any time soon.

Nevertheless the course is there for all to read, and when seen in the context of the medical lectures RS gave between 1920 and 1924, a very coherent story is possible to find. It may seem complex at first, but compared to other worlds views it is relatively simple (see 16, p. 238).

Part of its simplicity is, **there are only four energies, sourced from our astronomical background**, and they are the only players in every situation we will encounter in life. So keep an eye on them, and how RS names and describes them in different parts of manifestation.

The two discussions that can be seen are a) a discussion of how the energetic activities work **ONTO** each other, which is the 'Rings' story, and b) how they work **INTO** each other, which is the 'Arms' story of once things begin moving.

RS provides a 'Game of Life', which has the 4 activities manifesting in three interactive 'dimensions', as everything comes into manifestation. The dimensions are the **Cosmic** activities coming from the Rings diagram, and the **World and Internal** spheres, active in the Arms diagram. The Cosmic Activities sets a background field, within which the World Activities express themselves, and we 'internalised life forms' respond, with our internalised bodies, as best we can. Each of these dimensions manifest through a process of movement over a long period of time, which we can see as three distinct Stages, with three distinct sets of rules. Each stage is its own game, but functioning in the same space and time, as the other two.

Because we exist within several layers of spinning spheres, pretty much everything we experience, will be coming from their activity. So in practice, the Cosmic Activities and everything else we receive from above, externally can be considered as World Activities. Reality becomes the interplay of force activities from above, interplaying with our internal activity 'bodies'. Dr Steiner's natural science is the study of this very real phenomena.

The first step though is to pull the two conversations of the course apart. This is available in my book 'The Energetic Activities' , 2012. (16)

The First Conversation – Onto Each Other

The first story is of how the activities work according to the Cosmic Rings activity. Here they push and pull each other, in a dynamic interplay of the four primary energetic bodies.

One image RS gave of these activities was in his September 1924 lectures entitled 'Pastoral Medicine'. In this series of lectures he talks of what happens to ones inner experiences when any of the bodies are not in the right relationship to the others. He was talking to both priests and doctors, so these are wonderful stories telling how one or other of these disciplines will observe the same energetic imbalance.

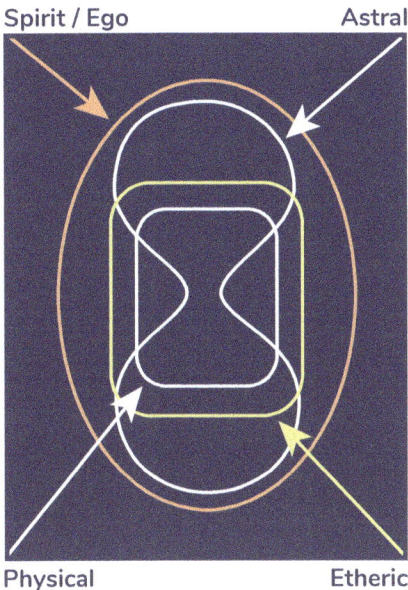

Here the white square is the physical body, while the yellow square near it is the etheric body. The etheric body extends a small way beyond the physical and acts as a 'watery cushion', holding off the astral - wavey white line, indicating movement - and the Spirit, indicated by the red circle.

The Astral body has this curvy shape, as RS is illustrating a dynamically moving process. Wherever activation is needed anywhere in the organism it has to come from the Astral body. The Spirit activity, while experienced as a contractive process, tends to show itself as an ordering and directing influence.

An analogy for these bodies activity is provided by the participants in the building of a house. The Spirit is the architect, who has the overall plan and provides the direction of the process, to follow for the overall project to be a success. These plans however have to be handed to the master builder, who is the Astrality. He provides all the motivation and necessary experience and skills to bring the plans into fruition. The Etheric body provides all the life-forces and energy of the workers doing the job, who use the Physical bodies substances, such as wood and nails to finish the job. Without the architect the builder will direct his energies in a unstructured and chaotic manner, so the building will not have the integrity it needs to 'be within the law' of life. If the Spirit is not present, the other activities loose focus, and do not do what they need to be doing. If the Astral influences are weak, then there is not enough basic force to stimulate the Etheric workers into action.

41

When we look back to the Astronomical reality from which these 'bodies' are sourced, the planets are essentially the only moving part of the game. The Stars are called Fixed Stars because from our perspective they do not move. We know that they are moving, but within our lifetime and indeed a 1000 years, they move only a very little distance, from our perspective. When we come to the Earths atmosphere and Etherics carrier, water, we can say that if it was not for the movement of the other planets, and their alteration of the EM field of the Solar system, the Earths atmosphere would be stable. Similarly, when water is left to its own devices, it will be calm and eventually stagnant. Water gains its life giving quality only when oxygen is wound into it by outside movement.

Hence the planets, and their manifestation as Astrality, are the only moving part of our reality, and we need it, the astrality, to be active for all the other processes, to have the motivation they need to carry out their tasks.

Any 'changes' in any of these bodies relationships will upset the working of the others. If the Spirit is too strong, then obsessive behaviour manifests, if too weak, one become spontaneous, scattered and often fearful. If the Astrality is too strong, it pushes off the Spirit and consumes the etheric. Thus manic depressive behaviour appears. Firstly due to the uncontrolled astrality bouncing around, and then once the etheric body has been consumed, the astrality invades the physical body too strongly and neurosis become 'stuck' and eventually the individual physically collapses. It can not escape the negative impressions of the astrality, as it is 'locked' into the physical. A less dramatic expression of this is artistic inspirations and mystical experiences, which are however hard to hold onto or do anything practical with.

If the etheric is weak, then illness of many kinds result, from a inhibited immune system. A weak Etheric allows the Astral to incarnate too strongly ,which leads to increased psychological stress and tiredness, mentioned above. If the Etheric is too strong, we only need to look to a pregnant woman for an example. The Spirit and Astral are pushed off, and so a pleasant scatteredness develops, while one puts on extra weight. If the physical body becomes too strong an array of sclerotic illness appear. If the other bodies do not connect with the physical body then one enters a stupor state, unconsciousness or coma.

This interplay of bodies creates a macro environment within which the second conversation takes place. This approach is very similar to the ancient system of the '4 humors', and many illness states can be resolved, by re aligning these 4 activities to their 'natural' state. The Biodynamic preparations work primarily upon these relationships.

The Second Conversation – Into Each Other

The second major conversation in the course, looks at what happens when these primary activities, all work more deeply into the physical sphere, to create the Physical Formative Forces. This story is best imaged in the picture given in the eighth lecture.

nerve-sense
system

metabolic-limb
system

earthly
substance

cosmic
substance

cosmic forces

earthly forces

RS terminology here has given many people great difficulties, so this whole conversation is one of those 'set aside'. Given around 25% of the text relates to this picture, it seems a huge step in the wrong direction to NOT understand and work with it. The Spirit activity becomes the Cosmic Forces, the Astral the Cosmic Substance, the Etheric becomes the Earthly Forces while the Physical becomes the Earthly Substance. These are activities standing just between the Physical body and the Etheric Body.

To address these activities properly though, we need to explore the whole reality of the bodies working into each others spheres, and then come back to this specific part of the story. So we need a frame of reference to address all the possibilities of all the bodies interworkings. RS did not address this question directly, in the Agriculture Course, however he did use different words for the different levels of activity. Sadly he did not make a glossary of terms to show how they all related to one another.

Finding a Glossary of Terms

In creation and in lifeforms, the Spirit works into the Astral, Etheric and Physical spheres. Likewise the Astral works into the Etheric and Physical spheres, and so on. Luckily these activities leave some footprints, and this is the organisational number of their activity. If the astrality works into the etheric it will leave a pattern of 7, (see the lectures on the 7 life processes), if it is the spirit it will be a footprint of 12 parts (see the 12 senses) and so on.

Thus RS's worldview is multi-layered and interactive with a mass of interrelationships to consider. 'Biodynamics Decoded' was my first attempt at organising all the various parts, via the vortex form, however it does not define the finer details very well. As a way of showing these more complex relationships I moved onto looking at the gyroscopic diagram, made up of both the reference systems - the rings diagram and the arms diagram.

The first diagram, is a cross section of the Rings pattern of the external gyroscopic beings, we live within. This archetypal pattern is repeated many times, in creation, as the center changes. Here it is the Earth, as this is where we are, and the rings are 'like' our EM field, as it is the center of OUR Universe. However we can put the Sun there, and the rings are those of the planetary spheres, or the center of the Galaxy, even nucleus of an atom, with its electrons. In each case there are spheres of activity around the center. This is a picture of all the major spheres of our experience, and called the 'Cosmic Spheres'.

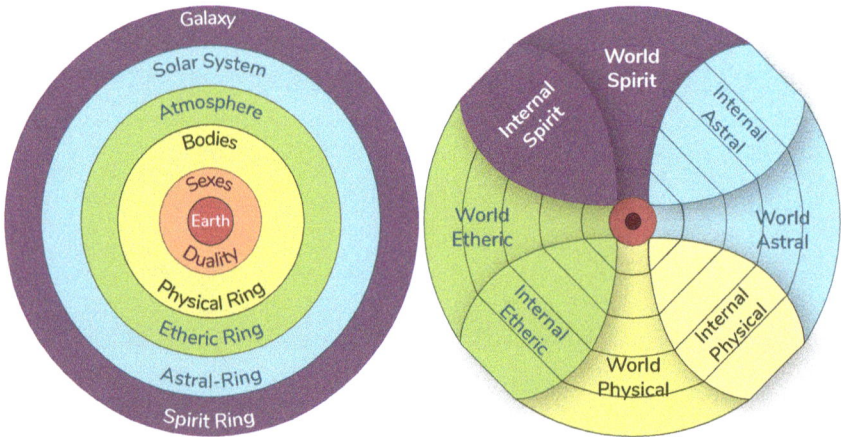

The second reference system, is when things start to move and polarisation occurs, leading to a gyroscopic form. This diagram provides two internal references. Firstly the 'primary cross' arms, show the internal influences of the World Spheres. This is a different relationship of the external bodies, to that shown in the Rings diagram, which is a pushing and pulling interaction on one on the other. This one shows how the external spheres work upon each other from an incarnated perspective through their polarity relationships. The second relationships shown in this diagram is the 'diagonal cross' arms, which are the zones of the internalised energetic bodies, which provide specific areas for the relationships we find within life forms. Both sets of arms are interactive with each other.

RS's stories, and his peculiar language is trying to convey the relationships between all these activities. So in every case he is trying to tell us where the Spirit, Astral, Etheric or Physical activities are working onto or into the various layers or dimensions of manifest life we are confronted with. So the question arises 'how can this be expressed in one diagram'?

If **we take the 'the Rings' and 'the Arms' diagrams**, both of which describe one type of energetic activity organisation, **and overlay them upon each other**, there arises an image of the activities working into each other. Take a few moments to put these pictures together.

I have done my best, using colour to indicate the activities of the two diagrams, however a little bit of imagination is still needed. The internal body's 'petals', I have left as whole colours, while on the external arms, I have shown the external rings. In an ideal world, these rings would be shown in the internal body zones as well, but then the 'strength' of indication for the internal bodies would be lost. The straight lines are used to indicate these rings acting in the petals.

This enlarged segment of the final diagram (p. 46), shows the complex relationships indicated by this interaction. On the horizontal World Etheric axis we have the purple area, being the Cosmic (and World) Spirit working into the World Etheric Arm, The blue is the Cosmic Astrality working into the World Etheric, the green has two Etheric references and is thus the World Etheric itself, with the yellow being where the Cosmic Physical works into the World Etheric. The Orange area is where the World Etheric supports the realms of exteriorised duality, which is a sphere I call the Earthly Substances. These are the basic substances from which everything arises.

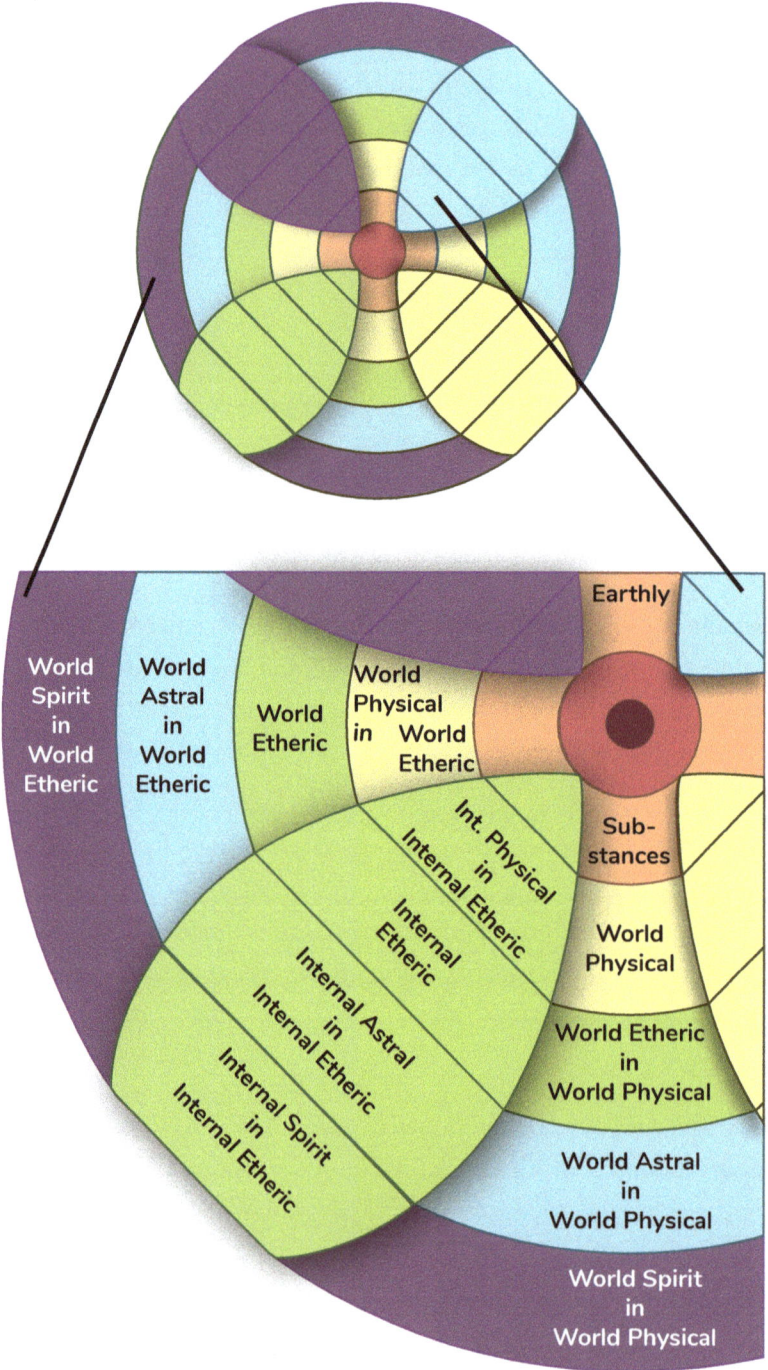

In the Green petal, the same process is indicated. The outer ring could be purple and indicates how the Cosmic Spirit stimulates the Internalised Spirit to work into the Internalised Etheric and so on. The complete chart of the energetic relationships is on the next page.

Essentially this diagram provides a basis for a **glossary of terms** used by RS, but possibly more importantly it also provides an energetic interpretation of all the 'double cross' monuments dotted about the planet. In many cases when you look closely at the more complex versions, such as the Tibetan mandalas and the many Persian and Christian designs, there are six layers of forms able to be seen, as well as the double cross. In addition to this in many of the Persian designs these rings are distanced according to the Golden Mean ration, to each other.

So while this organisation has arisen from a purely Goethean 'artistic' process, I have come to appreciate that it does have a high degree of 'truth'. One of my basic premises has always been that if we follow the 'As Above, So Below' doctrine presented throughout the ages, and stick with 'what is real', as the model for the steps taken, then what appears along the way should contain 'truth', that is at least worth investigating. This was certainly the case in my 10 year journey through "Biodynamics Decoded". I can say that while this gyroscopic form is very useful for understanding RS language, – as we will see shortly – it has also been very accurate in defining the activity of the chemical elements, and the activity one finds in the various parts of earth monuments, I have visited. Indeed I have found this organisation has opened up a effective form of healing, through the realigning of the bodies relationships, simply by moving around the octagon. This 'thought' is not new, as the Labyrinth movement can a test. The Chartres cathedral labyrinth was first made by the Knights Templar, in the 1300s, as a metaphorical journey akin to the journey of the crusades. Many people experience strong inner challenges as they walk the labyrinth, however there is no set interpretation of this journey. Until now... see the later chapter " The Chartres Labyrinth".

A key part of experiencing this organisations usefulness is one needs to have an experiential feeling for what the energetic bodies are, and how they feel when they are moved around with regard to each other. Simply by spending time with this chart and an octagon will provide the experience. Over a relatively short period of time the individual bodies will become tangible activities. You are made up of them, and they will move about in the octagon, so it is just a matter of time and perceptive sensitivity, before they become as real as feeling your foot step on a rock.

Glossary picture

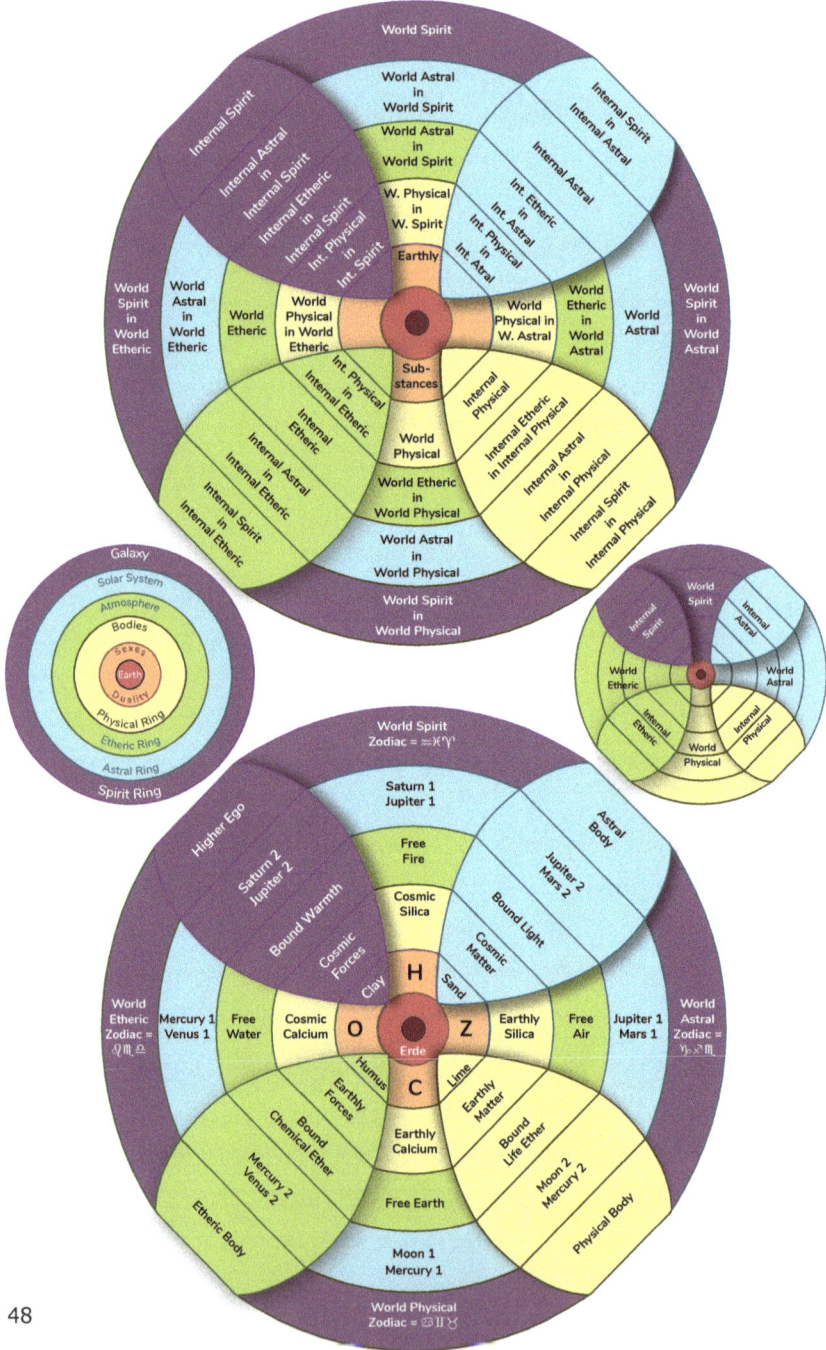

While we are on this subject, of moving the energetic bodies about, a very important tool in my 'awakening to the bodies' has been through using my **human essences**, and in particular my Willness, Astral Coolers and Etherics 3. These homeopathic biodynamic preparations work directly upon the energetic bodies interaction, just as RS said they would. The Willness, pull in the Spirit to stimulate a sense of personal control and direction, mostly by containing an overly active Astrality. The Astral Coolers, calm the rampant astrality, which is the cause of many different ailments, from pain, to emotional and psychological disruptions, to various forms of physical illness. The Etherics 3 stimulate the Etheric life bringing body, which is generally depleted by overly active Astral and Spirit goings on, along with the bad living habits they tend to bring along with them.

In chemical terms the etheric oxygen is bound up by the astrals nitrogen and the spirits hydrogen, thus the etheric cushion becomes depleted, so the astral and spirit can enter more closely into the physical organism, creating their havoc through illness psychosis and obsessive behaviors. CH is Methane, CN is Cyanide, need I say more. So if one can pump up the etheric and oxygen and push them off, while pulling off the astrality and if need be pulling in the Spirit, to help organise the rampant astrality, then most ailments can be 'adjusted' very quickly. The side effects of these processes are that you get to **experience the movement of the bodies** in relationship to one another. Sadly the essences are not strong enough so that in the early stages of 'treatment', one dose does not effect a permanent solution. On the positive side this then allows you to experience the bodies moving back again, if you stop using the drops. Which in turn allows for a further experience of the bodies moving again when they are again taken. In this manner the bodies become real experiences, and before long it is possible to know what feeling or experience of 'weirdness' is caused by what body relationship. At this point the essences can be taken just the once to effect the desired change. Having clarified these experiences before I began wandering the octagons helped a lot.

Naturally I contend, **this direct experience of the bodies, is essential** for anyone serious about working with Biodynamic Agriculture, or human health, as it takes one out of the head and theory, and into practical reality and realistic applications, of what RS was talking about. **It IS all about the energetic bodies...**

The diagram at the bottom of page 48, is the result of placing all the parts of creation Dr Steiner talked about in his 'Agriculture Course', onto the external and internal cross axis. The appropriate references and names he used in the course are placed in their appropriate energetic position. Each of the areas in the diagram are the carriers of a particular energetic activity, in a particular zone. E.g. The Green external ring has places for the 'free' or external atmospheric elements (and their accompanying ethers) for Water Earth, Air and Warmth, while the internalised areas on this same circle is the position of the 'bound' or internalised chemical, Life, Light and Warmth ethers and elements.

The 'Free Earth' is on the etheric ring of the external physical arm. This is the place where the Cosmic Etheric works into the World Physical sphere. This could manifest as water running through the soil and the activation of soil texture conditions, which other life processes may exploit.

The Free Water is the Cosmic Etheric ring of the World Etheric arm, so this is the most purely Etheric aspect of the gyroscope, and best expressed as the Water on the planet, and the life processes this facilitate. The Bound Chemical Ether is in the area of the Cosmic Etheric works upon the internalised Etheric. This would be a position that strongly supports the life and growth processes in any living being. It would be a significant point for the basis of very good health.

And so on. Each position on the gyroscope can be identified in this manner.

This diagram provides two significant bodies of information to cross reference against each other. The top picture is the energetic bodies interactions, and the bottom picture is Steiner's thesis on how these energetic activities interact in the kingdoms of nature. The combination of these two systems is revealing, and imaginatively useful, to comprehend Dr Steiner's schema, as well as his practical suggestions. This effectively provides the missing glossary of terms, for the Agriculture Course.

Most importantly, this complex of information can also be used as a basis for further cross references with anything that fits this same patterning. The latter part of this book explores the reorganisation of the Periodic Table of Chemical Elements according to this pattern.

The Agriculture Course

To be able to grasp some of the applications of chemistry to life, we need to look further into the second conversation in the Agriculture Course, to gain a useful practical framework. This is the best original diagram for this conversation. Sadly it only appears in the last lecture rather than in lecture 2. In it there are two reference schema. On the top line we have the organism divided in the nerve sense / head region and the metabolic-limb / belly region. In the first few pages of lecture two, RS describes how these two polar activities work towards each other to form a central mutable zone that shows as our rhythmic system, with its lungs and heart processes. This threefold order is the basic structure found in the physical bodies of most of the higher life forms.

This piece of the story is generally 'appreciated' by most BD storytellers, however 'the problem' arises when they stop there, and try to understand the rest of RS story, from this layer alone.

Below this threefold layer, each of the poles is divided into two activities, with the obscure names Cosmic Forces and Earthly Substance, and the Earthly Forces and Cosmic Substance.

RS is telling us here that **these four processes are taking place inside the threefold physical organism**, and if we really wish to control the growth of anything, we need to work with these processes. So what are these processes and how do they relate to the four primary energetic activities?

The key to this question can be found in a passage in lecture 6 of 'The Healing Process, given on 16.11.23 *"Although the three major systems intermingle, they are distinctly different from one another. The physical, etheric, astral and*

I-organisations work totally differently in our sensory nervous system, for example than they do in our rhythmic or metabolic-limb systems. All four members of the human constitution – physical, etheric, astral and I– are present in each of the three spatially somewhat separate systems, but affect each one in very different ways." So the four 'Cosmic and Earthly' terms used in the Agriculture course, are RS way of indicating how the 4 energetic activities manifest at this physical level, which 'spatially' manifests as a threefold structured body. When we work into this further, instead of telling us about the 4 activities in each sector, he is focusing upon the dominant 2 players in each sector.

The Physical Formative Processes

But what of the physical processes? From the bottom diagram on page 48, we can see the terms in question are all on the cosmic physical ring. Cosmic Forces appears in the Internal Spirit arm, Cosmic Substance is on the Internal Astral arm, Earthly Forces is on the Internal Etheric arm, while Earthly Substance is on the Internal Physical arm. Thus these are the dominant sources of the processes RS is describing with those words. Whenever one sees the word 'Cosmic Substance', you should think to yourself the Astral body is working into the Physical. Similarly if you see 'Light Ether' you should think the Astral is working into the Etheric body, and so on.

Dr Steiner's medical lectures clarifies this a bit further. His descriptions there suggest with the words 'Cosmic Substance' he is really indicating the working of both the Astral and Spirit in the metabolic system, however the Astral is the dominant player, with the Spirit playing a secondary role. Similarly 'Cosmic Forces' is really saying the Spirit and Astral activity within the nerve sense system, however the Spirit is the dominant influence, while the Astral is the secondary player within this system. The same is true of the Earthly processes. Physical dominates Etheric in the nerve sense system while the Etheric dominates the Physical in the metabolic system.

So, in the PFF story we find the parts working as polarities in the physical zones. The Spirit and the Physical activities of the vertical axis of the gyroscope polarise, and work predominantly together in the nerve sense system, while the Etheric and Astral activities of the horizontal plane also work as a predominant creative polarity, in the metabolic system. Both these joint sets of activities work towards each other, which leads to the creation and the ongoing sustenance of the Rhythmic system, that is made up of the lungs and the heart. The lungs process is a 'manifestation of the head working into the middle zone, while the heart and circulation is a manifestation of the metabolic processes working upwards. Health is were these two major activities are working in their proper manner.

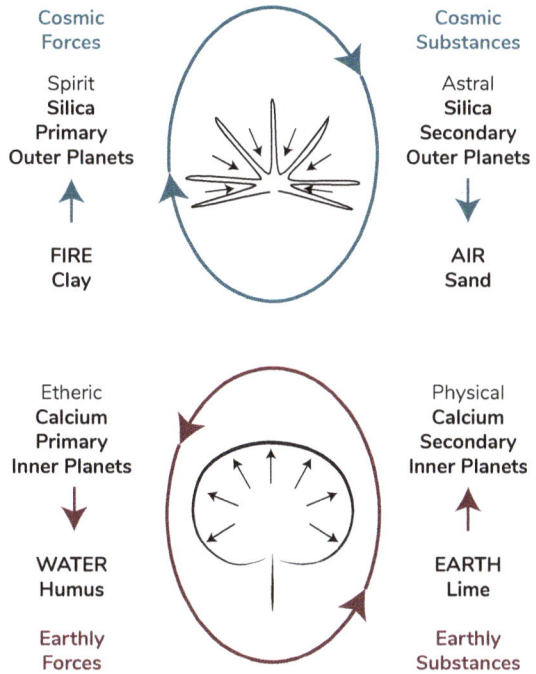

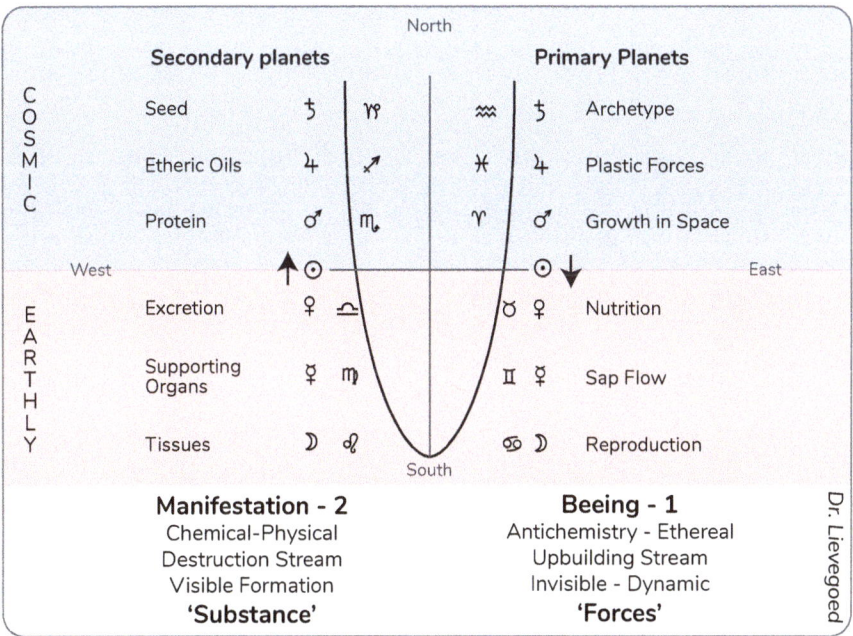

Cosmic Forces

Spirit
Silica
Primary
Outer Planets

FIRE
Clay

Cosmic Substances

Astral
Silica
Secondary
Outer Planets

AIR
Sand

Etheric
Calcium
Primary
Inner Planets

WATER
Humus

Earthly Forces

Physical
Calcium
Secondary
Inner Planets

EARTH
Lime

Earthly Substances

	Secondary planets				Primary Planets	
COSMIC	Seed	♄	♑	♒	♄	Archetype
	Etheric Oils	♃	♐	♓	♃	Plastic Forces
	Protein	♂	♏	♈	♂	Growth in Space
West		↑☉		☉↓		East
EARTHLY	Excretion	♀	♎	♉ ♀		Nutrition
	Supporting Organs	☿	♍	♊ ☿		Sap Flow
	Tissues	☽	♌	♋ ☽		Reproduction

North / South

Manifestation - 2
Chemical-Physical
Destruction Stream
Visible Formation
'Substance'

Beeing - 1
Antichemistry - Ethereal
Upbuilding Stream
Invisible - Dynamic
'Forces'

Dr. Lievegoed

The top picture on p. 54 provides an image of RS story. It also includes the arrangement of the activities in the external environment, as well as the internal activities RS showed in the lecture 8 diagram. The external activities also need to be also considered, if we are to keep the story straight. Externally the Spirit and Astral activities come from above, and work onto the head region of the animal, while the Physical and Etheric processes come from the Earth 'upwards'. (lecture 1). RS talks of these activities also in lecture 4 of the course, however they are not clearly defined from the internal processes, which polarise. The internal activites are discussed later in the same lecture, when he talks of how the Etheric and Astral work together in the metabolism to create manure.

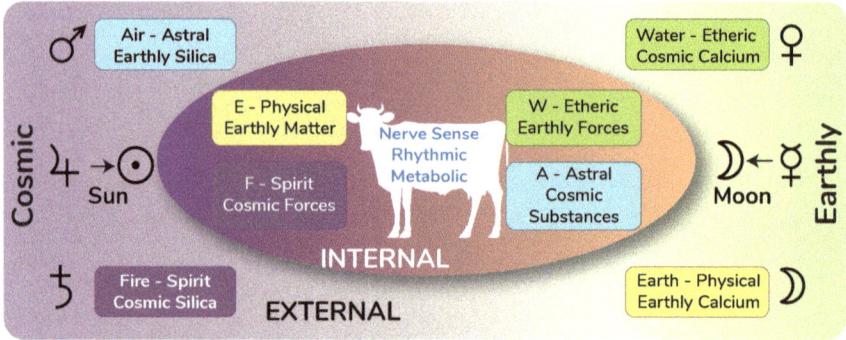

The 'oval' is the internal organisation of activities that sits inside of the external organisation. The plant has a slightly altered arrangement as the internal organisation of the plant is reversed to that of the human and animals, with regard the external environment. The root zone is the nerve sense area of the plant, while the metabolic zone, which has all the reproductive organs is the flowers and seed zone, thus its belly is facing the sky and its head is towards the centre of the Earth, opposite to the human. So it adds a bit of complication (16).

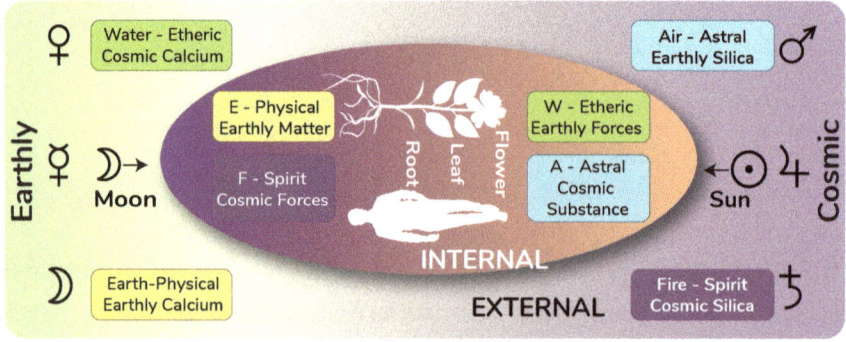

RS is very specific with the details he provides of how and where each of these poles work and interact. This is all contained within the medical lectures given in 1920 and 1921, which can be obtained on the web at http://www.rsarchive.org/Medicine/, for free. As much as I would love to quote endless passages from these works, this publication is not the place. If you are serious about this study you need to read these lectures anyway. They are free and available so you may as well go to the 'horses mouth'.

RS's story begins, with the two opposite vortexes, the left hand spiral from above and right hand spiral from below, working towards each other. In the second lecture of 1920, he says *'The polarity in man is only comprehensible if we know that his structure is a dual one and that the upper portion perceives the lower. The following too must be considered: the lower functions – one pole of the whole human being – are considered through the study of nutrition and digestion in the widest sense, up to their interaction with respiration. The interaction goes on in a rhythmic activity; we shall have to consider the significance of our rhythmic system later. But linked up with and belonging to the respiratory activity there is the sensory and nervous activity, which includes all that appertains to external perception and its continuation and its being worked up in the nervous activity. Thus, respiration and sensory and nervous activity form one pole of the human organism. Nutrition, digestion, and metabolism in its usual sense, form the other pole of our organisation. The heart is primarily that organ whose perceptible motion expresses the equilibrium between the upper and lower processes; in relation to the soul (or perhaps more accurately in the sub-conscious) it is the perceptive organ that mediates between these two poles of the total human organisation. Anatomy, physiology, biology can all be studied in the light of this principle; and thus light is thrown, and only thus, upon the human organisation. As long as you do not differentiate between these two poles, superior and inferior, and their mediator the heart, you will not be able to understand man, for there is a fundamental difference between the two groups of functional activity in man, according to whether they pertain to the upper or the lower polarity.*

The difference amounts to this: all the processes of the lower sphere have their "negative" so to speak, their negative counter-image in the upper. The important point, however, is that there is no material connection between these upper and lower spheres, but a correspondence. The correspondence must be correctly apprehended. without search for or insistence on direct material connection.'

These two spheres have their own characters and ways of working in lecture 7 (p. 105, 1920) RS described them this way – *'Thus man is affected in the most diverse ways, by telluric forces (Earthly Substance / Physical) – call them terrestrial if you so prefer – and by extra-telluric forces.(Cosmic Forces / Spirit) If we wish to study these forces, we must look at the result of their co-operation in the whole human entity. They cannot be traced in any isolated part of man, and least of all in the cell – please note especially, in the cell least of all. For what is the cell? It is the element that obstinately maintains its separate existence, its own separate (Etheric dominated) life and growth, contrary to the whole of human life and growth. Picture to yourselves, on the one hand, man built up in his whole frame by the telluric and extra-telluric forces, and on the other hand, the cell as that element which intervenes in the operation of these forces, upsets their ground-plan and conception, and even destroys their working by developing its own urge towards independent life. Actually we wage a ceaseless war in our organism against the life of the cell. And the most impossible of conceptions has just arisen in that Cellular Pathology and Cellular Physiology which find cells as the source and basis of everything, and regard the human organism as an aggregate of cells. Whereas, in truth, man is a whole in relationship with the cosmos, and has to wage perpetual war against the independent life and growth of the cells.*

In fact the cell is the ceaselessly irritating and disturbing factor in our organism, not the unit of construction. And if such fundamental errors enter into the general scientific view, it is not to be wondered at if the most mistaken conclusions are drawn regarding the nature of man in all its implications.

So we may say that the contractive formative process of man and the expansive process of cell formation represent, as it were, two opposite sets of forces. The individual organs are right amidst the action of these forces; they become liver or heart and so forth according to whether the one or the other set of forces prevails. They represent a continuous balance between two poles. Some of the organs tend towards the cellular principle, and the cosmic factors have to counteract this tendency. Or again, in other organs – which we shall presently specify – the cosmic action dominates the cellular principle. In the light of this knowledge, it is especially interesting to observe all the organic groups that lie between the genital tract and the excretory tract on the one hand, and the heart on the other. These organs, more than any others, resemble the actual state towards which cellular life tends to develop. This resemblance is noticeable in comparison with all the other organs of man. *And we must draw the following conclusion as to the essence of the cell. The cell develops – let us exaggerate somewhat, but consciously, and in order to make our point clear – an obstinate and antagonistic life, a life of self-assertion. This obstinate life centered in one point meets the resistance of another force, external to it. And this external element counteracting the cellular process, takes away the vitality from its formative forces. It leaves untouched its globular shape as of a drop of liquid, but sucks the life from it, as it were.*

This should be an elementary piece of knowledge familiar to all; everything on our earth that is globular in form, whether within or external to the human frame, is the result of the interplay of two forces, one urging towards life, the other drawing life away.

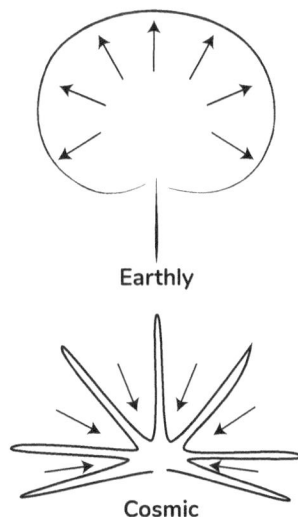

Earthly

Cosmic

If we examine the concept of the mercurial (middle process) in ancient medicine we learn that it was held that the mercurial has been deprived of life but retains the globular form. This means that the mercurial element must be visualised as tending obstinately to the condition of a living drop of matter, i.e., to a cell, but as prevented by the planetary action of Mercury from being more than a corpse of a cell – that is to say, the typical quicksilver globule.'

In this passage we are given images of the cellular process as a expansive process, and the cosmic process as a contractive containing process. In Agriculture this is the same conversation as that between the two processes of Earthly Calcium being the expansive influence, while the two Cosmic Silica processes are the contractive.

These processes not only interact with each other – as in push and pull – but they work right into each others spheres of activity. The Spirit and Astral, while centered in the head region, work right through and into the metabolic region. Similarly the Physical and Etheric processes, while centered in the metabolic region, work right through the organism and up into the head zone. In a lecture given only weeks following the Agriculture course RS says – *'Thus when we explore the wonderful internal structure of the human organ-*

ism, we discover not only a generative and regenerative process in each individual organ, an activity that serves the organ's growth and continued development, but also a degenerative process that reverses physical development, but makes it possible for the soul spiritual element to find its place within the human being. I said last time that the specific balance between regeneration and degeneration in each human organ can become disturbed. When regeneration becomes over whelming, inflammation disease conditions result.' (p. 117 , Healing Process 21.7. 1924)

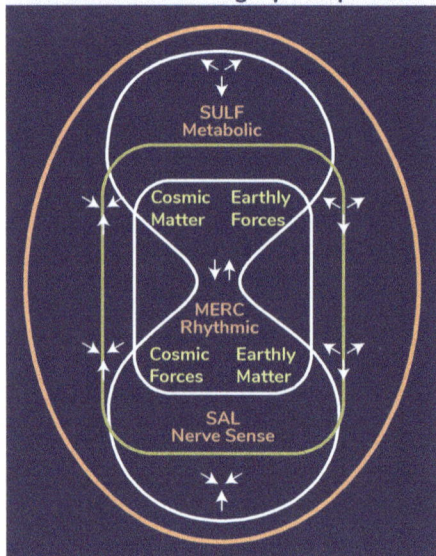

Therefore within both of these regions, the head and metabolism, a 'struggle' exits between an expansive process, coming from the belly, and a contractive process coming from the head. The head processes have a hardening devitalising effect ,while the belly processes show as a softening or over vitalising effect. Illness, in RS eyes, is where one or other of these processes becomes either too strong or too weak within any particular area.

In RS language, the head region with its relationship to Silica works primarily through Cosmic processes, but has also a secondary Earthly process coming from the metabolism. While the belly region and its relationship to Calcium works primarily through the Earthly processes, however with a secondary cosmic process coming from the nervous system.

In the Head we need to work with both a contractive and expansive activity present there. The same is true of the Belly. This basic 4 fold activity, was imaged by the Chinese many years ago as the yin yang symbol. In between these stark polarities stands 'the middle' processes that arise out of these two poles interactions. Therefore we can also look for the 'Mercury' processes between the 'Salt' and 'Sulphur' polarities (see p. 162).

To clarify the Cosmic and Earthly activities further, we can refer to lecture 2 of 'Agriculture', where RS talks of how the energetic activities, use the Calcium and Silica processes to do their bidding within the physical organism. He talked of how the Silica processes have two sides to their action. One coming from above, which is firstly 'absorbed' by the Earth, before this activity is then radiated back outwards through the plant growth and other life forms.

The Silica Cosmic process 'starts' with the ripening nutritive activities of Light, which carries the Astral 'Cosmic Substance' activity, and Warmth, which carries the Spirit 'Cosmic Forces' activity, being drawn into the soil during the Autumn. The light process is held by the Silica sand in the soil, and in the spring with the help of clay, the warmth based Cosmic Force is pushed back upwards through the plant. This carries the impulse to set viable seed for the future generation. For fruit and seed to develop properly, there needs to be an active interplay between the Cosmic Substance activity, with the rising Cosmic force activity, coming from below. If this contact is not made properly, then the plant will flower, however fungal attack will destroy the flowers and fruit, due to there being insufficient upward force to push the plant all the way through to seed ripening. In this case, RS comments in the 6th lecture, that it is the Earthly processes which dominate the Cosmic Forces, and fungus results. If the Cosmic Substance process is too strong then insect attacks are very likely, as the softening Earthly processes can not cope with this inward push of Astrality.

The 'Earthly' Calcium / Physical processes are best seen as starting from the soil, where water and minerals are drawn into the plant, and begin to spiral upwards through the plants channels and veins. This activity is central to the quality of the tissue formation. The veins become increasingly small, which helps for the minerals to be taken from the water. Eventually the water, which has now been significantly potentised from all the spiraling, is puffed out into the atmosphere as transpiration. This 'sensitive water' becomes the moisture in the atmosphere, which RS called 'atmospheric calcium', and Earthly Forces. You could also say it is the World Etheric working into the Physical body. This refined moisture is either drawn back into the plant, or falls to the Earth as dew. RS commented that the humus in the soil acts as a attractive 'trap' for this living activity from the atmosphere. The Calcium process in the Earth helps with germination of the plant and the quality of its tissue formation, which is beneficial to plant health and keeping quality, and is called Earthly Substance. While the atmospheric calcium, which helps with supporting fertility and pollination, while helping with the enlargement of the plants parts, carries the etheric activity, and is called Earthly Forces.

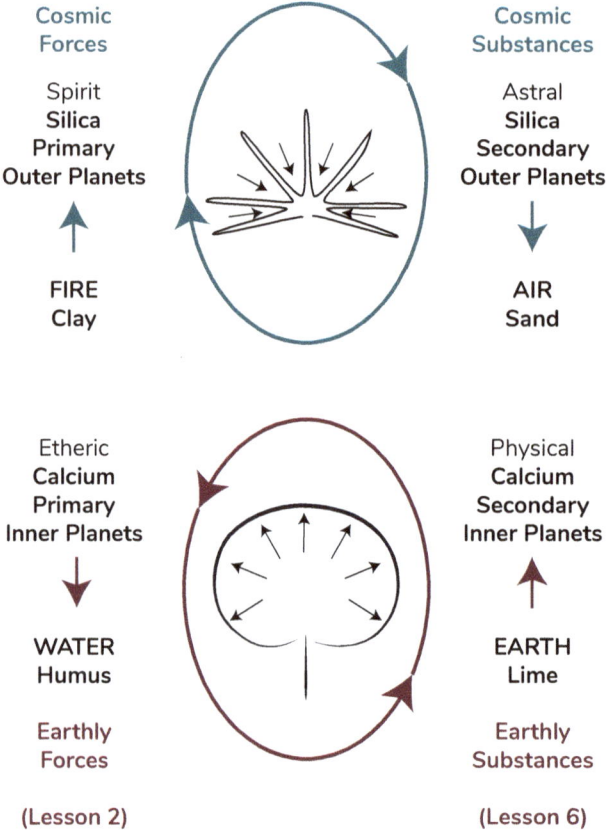

Cosmic Forces	Cosmic Substances
Spirit **Silica** Primary **Outer Planets**	Astral **Silica** Secondary **Outer Planets**
FIRE Clay	**AIR** Sand
Etheric **Calcium** Primary **Inner Planets**	Physical **Calcium** Secondary **Inner Planets**
WATER Humus	**EARTH** Lime
Earthly Forces	Earthly Substances
(Lesson 2)	(Lesson 6)

The message from this is that, the internal Cosmic and Earthly processes RS describes are not a two fold polarity, as they appear externally, but a 4 fold interaction. It is ONLY when this phenomena is understood that much of the Agriculture course makes sense.

A further consideration of the Cosmic and Earthly question is how these four substances Clay, Sand, Humus and Cations relate to the substances of Calcium and Silica. RS made very clear statements about these four substances in lecture two, however at the end of lecture three he made a very vague comment of 'clay being a mediating substance between Calcium and Silica, although it is closer to Silica'. This comment has given the 3 fold fanatics the idea that clay is THE mediating substance between the two, which means they completely deny the three lecture two comments that 'clay strengthens the upward moving silica process'. The solution to this 'problem' appears to be that RS should have said that the four substances and more so their processes, mentioned in lecture 2 , are all 'middle' processes taking place within the external Calcium / Earthly and Silica / Cosmic processes, he outlines in lecture 1. (The planets naturally follow on from this pattern.)

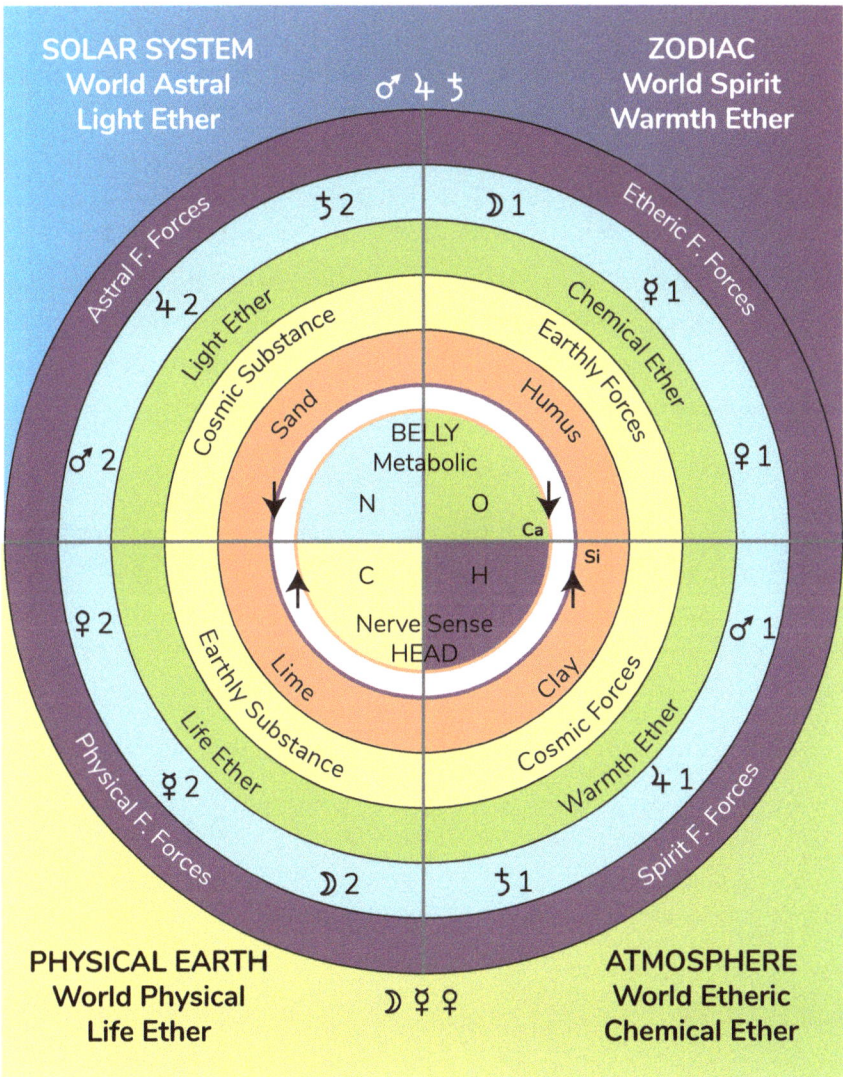

Interestingly, in lecture 2 of the 1921 medical lectures, there is a story of how the etheric body, and its parts, the Ethers, work to manifest as illness. In that story, the polarisation, which we find in the physical processes, does not appear. The Warmth /Spirit and Light / Astral works from the Head downwards, while the Life and Chemical ethers work from the Belly upwards, and illness is caused by their imbalance as described in the push / pull reference used for the primary bodies. This difference between how the Etheric Formative Forces work together and how the Physical Formative forces work together, is not understood by those who only acknowledge the Ethers as THE primary formative principles, and thus much of the Agriculture Course makes no sense to them. This fundamental difference, Ethers working as primary polarities, and the Physical Formative Forces (PFF) working as secondary polarities – see Biodynamics Decoded, needs to be bought to clear consciousness. Very little was said of the Ethers function in the Agriculture lectures, while the PFF are described in around 30% of the text. This is the context for much of what is said there. The PFF have a direct relationship back to the energetic bodies actions within living forms, and so working with them allows Biodynamics to be cross referenced with the medical lectures, and huge avenues of understanding and new exploration arise.

See "The Physical Formative Forces"

http://garudabd.org/wp-content/uploads/The-Phy-Form-Forces-23.pdf

The Gyroscopic Periodic Table

The Gyroscopic Periodic Table has been a natural progression developed from a study of Soil Science, Dr Steiner's Agriculture and Medical Courses, Astronomy and Astrology. I am still an early student in chemistry and do not pretend to have an extensive knowledge of it. I offer the accompanying diagrams and explanation, as the beginning of what I hope will become an ongoing discussion of both the questions and answers this diagram presents.

The process of coming to this diagram is best described as an artistic interpretation of the scientific evidence I have been able to reference. What is outlined here, is the outcome of the process I first outlined in "Biodynamics Decoded". One main aspect of that work was the identification of an archetypal patterning, standing behind life on Earth and creation in general.

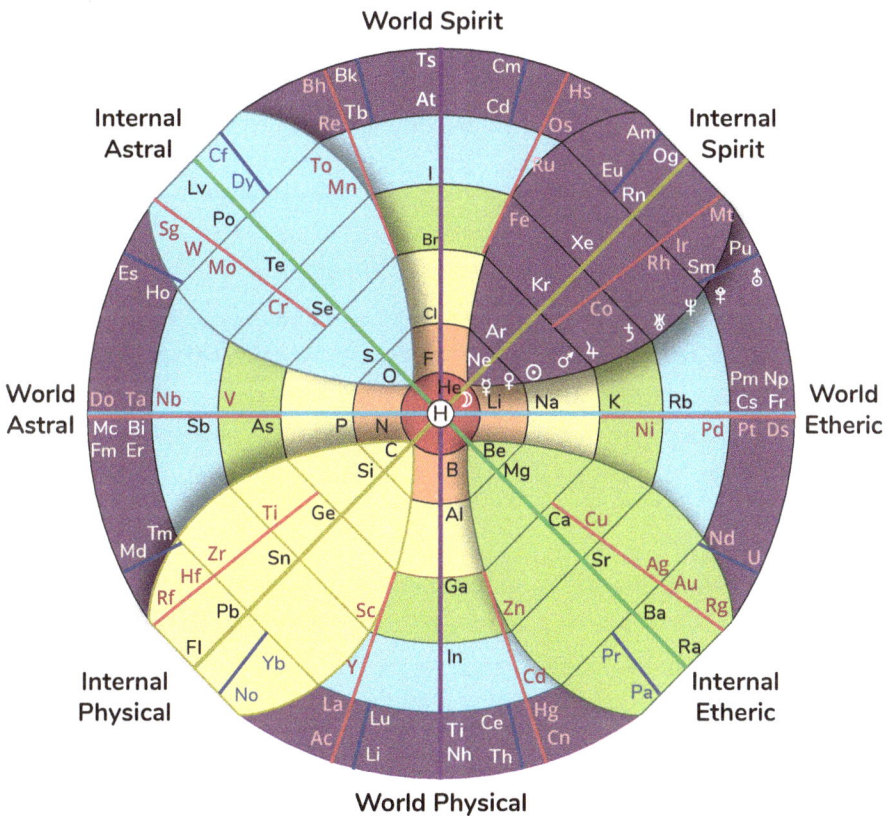

This thesis, I have come to call "the Atkinson Conjecture" (see Appendix 1) suggests, if we look at life processes according to the laws that arise from the

Astrological / Biodynamic Gyroscope, we will uncover 'truths', which at least, can act as starting points for questions, and practical trials, to show us how creation manifests. The thesis says the quality of the questions will be such that they will, more often than not, reveal actual truths one can work practically with. Because we are seeking 'archetypal truth' it would be expected that this patterning should be able to be seen everywhere.

The practical application of this thesis to date has been confined to using the Biodynamic preparations, as its functional tools. With the thesis's clarifications of Dr Steiner's suggestions, it provided many real and unique influences on plant growth and animal control. Three of these applications; ThermoMax – frost protection, BirdScare – bird control, and PhotoMax – photosynthesis enhancement, have gained third party scientific proof, (HortResearch NZ) which is available for review on the BdMax website. The successful commercialisation of these products speaks further to the usefulness, effectiveness and practical reality of 'the Atkinson Conjecture'. The question arose though as to how these preparations might become stronger in their action. The chemical elements became an obvious possibility.

The following pages, provide a study of the chemical elements, which science accepts as the basis of material manifestation, but seen through the wormhole of the Gyroscope. **Given the chemical elements are a natural manifestation of our environment, it stands to reason that they must organise according to the same gyroscopic principles, as everything else**. In the process outlined here, many new understandings of the chemical elements are revealed, while the 'old' information is organised in new ways, and cross referenced against everything else we already know about the archetypal gyroscope.

If nothing else, my organisational diagrams are a practical approach to objectify Dr Steiner's ideas into a pictorial form, hopefully making it easier to comprehend the holistic approach he suggests. Just as I found that Dr Steiner's world view uses the same basis as Astrology, so with the Periodic Table, it too has the same structural basis as the Gyroscopic Glenopathic world view, arising from "Biodynamics Decoded". It seemed a natural next step to join them all together. An act that allows **the energetic activity of every element to be 'conjectured'**.

I have not yet come across a work suggesting the spiritual activity of all the elements of the Periodic Table. So, the information this diagram suggests, must firstly be taken as indicative. It is a challenging question generator from the Universe and it provides suggestions, for how any particular element might work. My explorations to date, have convinced me of its usefulness, enough to believe it is worth making available to others, who might like to explore **the chemical realm, as influences of the energetic body interaction**.

One of these people is Hugh Lovel, who I have been bothering with 'my chemistry' since 2001. He recently sent the following comments through:

"You have my vote for making sense of the periodic table. I'm always following up and confirming your intellectual realizations. My hat's off to you, my friend. I owe you a great (and growing) debt for deeper understanding of biodynamics and beyond, which I hope including mention of your work in the tail end of my book, Quantum Agriculture, is a first payment for. Thank you , thank you, thank you. Best wishes, Hugh Lovel"

I put this here, as words of 'one who has bothered to look'. In the hope it will encourage you further to truly investigate this approach.

There are 120 elements in three groups. **The task is to identify each elements activity** and then apply this to nature, as an safe and pure methodology.

For the major elements I have given a 'thinner', quick, example of how the reference systems works. However, for the transition elements, I have provided deeper examples of the referencing process that can be used to identify an elements possible energetic activity. The same process of knowledge collection and reflection I show there, can be done for the major elements and Rare Earths. With the Rare Earths chapter, which was expanded in 2017, I have taken a similar process of information collection, however my interpretations of the Lanthanides are informed by my direct 'provings' of the elements, as well as 'the system' analysis of the accepted information about them.

The lemniscate process outlined for identifying the transition elements presents us with a good example of a 'wild call', that arises from following the processes identified with the 'Atkinson Conjecture', along with insights from Dr Hauschka.

It is with a spirit of exploration, and the desire for the worth of this work to be deepened and valued, or revealed as worthless, that I make it freely available to all who would like to take part in its unveiling.

The following pages are a start, and further detail is expected to be added as the journey continues. Please share your insights and questions, either directly with me at garuda@xtra.co.nz. See www.garudabd.org for the earlier books in this series, from which this book is an extension.

The Periodic Table

1	2	3	4	5	6	7	8	9	10	11	12	3	4	5	6	7	8
1	2	3	4	5	6	7	8	9	10	11	12	13	14	15	16	17	18
1 H																	2 He
3 Li	4 Be											5 B	6 C	7 N	8 O	9 F	10 Ne
11 Na	12 Mg											13 Al	14 Si	15 P	16 S	17 Cl	18 Ar
19 K	20 Ca	21 Sc	22 Ti	23 V	24 Cr	25 Mn	26 Fe	27 Co	28 Ni	29 Cu	30 Zn	31 Ga	32 Ge	33 As	34 Se	35 Br	36 Kr
37 Rb	38 Sr	39 Y	40 Zr	41 Nb	42 Mo	43 Tc	44 Ru	45 Rh	46 Pd	47 Ag	48 Cd	49 In	50 Sn	51 Sb	52 Te	53 I	54 Xe
55 Cs	56 Ba	57 La	72 Hf	73 Ta	74 W	75 Re	76 Os	77 Ir	78 Pt	79 Au	80 Hg	81 Ti	82 Pb	83 Bi	84 Po	85 At	86 Rn
87 Fr	88 Ra	89 Ac	104 Rf	105 Db	106 Sg	107 Bh	108 Hs	109 Mt	110 Ds	111 Rg	112 Cn	113 Nh	114 Fl	115 Mc	116 Lv	117 Ts	118 Og

58 Ce	59 Pr	60 Nd	61 Pm	62 Sm	63 Eu	64 Gd	65 Tb	66 Dy	67 Ho	68 Er	69 Tm	70 Yb	71 Lu
90 Th	91 Pa	92 U	93 Np	94 Pu	95 Am	96 Cm	97 Bk	98 Cf	99 Es	100 Fm	101 Md	102 No	103 Lr

This rectangular Periodic Table of elements is the form accepted by science, and by which the nature of the chemical elements are described. These elements make up the basis of all material forms, both living and dead.

The basis of this table is a series of columns listed along the top, today, numbered 1 to 18. The original periodic table developed by **Dimitri Mendeleev**, had the chemical elements divided into 8 primary groups, later the 10 groups of transition metals where classified between group 2 and group 3. Then later again the 14 groups of rare earth elements were added.

Each of the major group has 7 layers. So there are 8 primary groups or arms of major elements. Two have 7 members each, while the other 6 have 6 members each. The transition elements have 10 columns but only 4 layers of elements, while the Rare elements have 14 columns but only two rows of elements.

When using this table, we are asked to imagine for the actual structure of any particular element, a nucleus comprising a number of protons and neutrons, with rings or spherical shells of electrons around it. This is the same image as the image of the Solar system put forward by Copernicus. Each new layer of the normal table above, indicates that the elements in that layer have another shell of electrons, added to those of the layer before. So as we move up the table, starting with Hydrogen, we have a positively charged proton nucleus with one negatively charged electron in the first shell. For Helium, the next element, we have an extra proton and 2 neutrons with another electron in the first shell. Each subsequent element has an extra proton and neutron, in the nucleus, and another electron in the first shell. The next shell, begins with the 3rd element Lithium. Once that shell has its eight electrons, at Neon, the electrons begin to fill the next shell, with Sodium. And so it goes on through the chart.

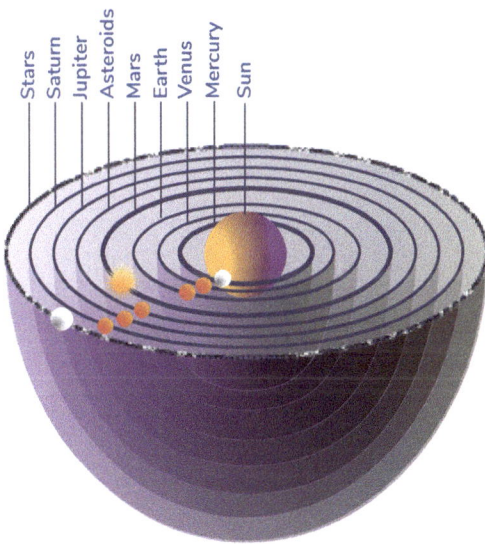

The interesting thing about the standard Periodic table is that **while it is represented as a rectangle, we are continually asked to imagine it as a spherical nucleus**, surrounded by spheres of electrons.

I found this imagination process became one of my earliest hurdles when I began to work with the Soil Science and the Periodic Table. So my first question was, **why not draw the Periodic Table as a circle**, if only for my own ease of picturing the reality of the elements.

Because the Table has eight major groups of elements, it is easy for the arms to be placed around the circle in an ordered manner. Once the Periodic table is circular then all the rules we have for circles, especially those coming from Astrology and the Biodynamic Gyroscope can be applied to the circle, if for no other reason than to establish an 'artistic' cross reference.

The main considerations that arise are: **the circle divides according to the gyroscopic form, with a vertical and horizontal axis**. Astrology identifies these primary axis as the Mid-heaven (vertical) and Ascendant (horizontal) is, which provides the cardinal angles upon which the 'houses', or areas of Earthly manifestation, are arranged. The **ascendant**, which is the eastern point at the time of birth, is the beginning of the circle of houses. This is the 'new individual' who begins their journey through life. Thus we have a **beginning point**. We can make the assumption from this that the arm of the first group which really starts with Lithium (as Hydrogen is the middle element) would be placed here.

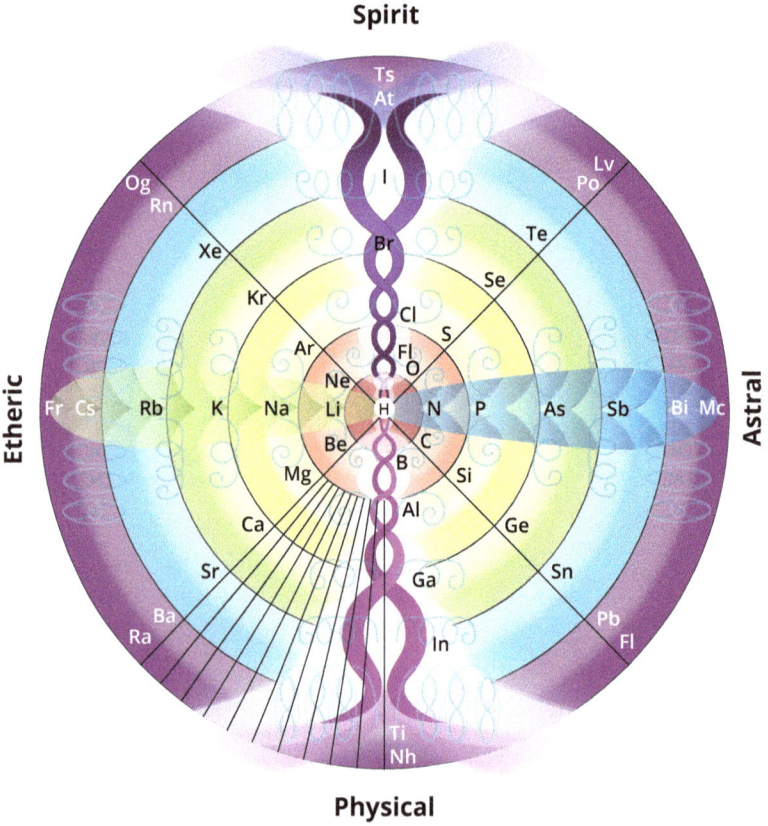

While this assumption can be made, I was not satisfied with this simple reference providing a truth, so I spent some years examining all the arms to see where they fit, and came up with Lithium at the ascendant nevertheless, for other reasons. Dr Hauschkas stories (18) of the elements were helpful in this process.

Within Astrology, polar relationships are identified to exist between opposite points. For example if the influence of Libra is predominant in a chart, then the opposite influence of Aries, must also be considered. Lievegoed showed the polar interaction of the planets working together in the manifestation of plant processes, and so on.

So in the case of the Periodic Table, we are presented with a series of **relationships between the arms, and the elements of those arms, which are not immediately obvious in the rectangular chart**. For example Magnesium and Sulphur (Epsom Salts), are polar elements, as is Calcium and Selenium; both elements are necessary for maintaining good health. In soil science the relationship between Potassium activity and Phosphorus is experienced by growers, yet very little research is done into their direct relationship to one another.

In soil science it is well known that 'related' elements act against each other. For example if Calcium is very high in concentration in the soil, then both Magnesium and Potassium will be 'weakened' in their activity. The question arises, what about the effect of the polarise element? What is occurring with the Selenium if the Calcium is high? What happens to potassium when either nitrogen or phosphorus is over applied.

While simple opposition relationships show up regardless of where the arms sit on the circular picture, the next question of the circular periodic table, is **where should the arms be placed in relationship to the Biodynamic gyroscope**. It could be considered of no concern, however due to the immense organisational similarities between the Periodic table and all of the information we have gathered so far, from the Biodynamic gyroscope, it would seem a reasonable question to ask, so that further cross references can be made.

The positioning I have chosen in this diagram is based upon many considerations. The more I work with it, – and the more I learn about the elements – the more I am happy with this choice. Apart from the obvious association of the first arm of elements , The Alkalis, with the beginning of the houses – on the ascendant – a couple of other significant considerations were, that the Halogen elements – Fluorine, Chlorine, Bromine etc – are very reactive and corrosive and very toxic. RS characterised the halogens as bringing processes to a halt. This is what the Spirit does when it is overly active. It kills things.

It burns and vaporises the life processes it touches. The arm to the left of the vertical Halogens is that of the inert gases, which do not change their character or react with anything around them. They maintain their individuality above all else, which is what the spirit can do when it works strongly within a personality. Indeed I have come to see that this is the arm of autism, where the individuals are very self contained and actually in no need of the world.

Other arms such as that with Calcium and Magnesium, two elements that make things bigger, become the carriers of the internalise growth processes, seen with the internal etheric. While the Carbon Silica arm become the carriers of the internalise physical processes. Thus the major arms are placed.

Now that we have all the parts of this 'puzzle' in place there are several layers of cross reference that can be bought to bear on the chemical elements.

The cross-reference material for the Periodic Table can be outlined as:

A) **The chemical understandings of common chemistry**

B) **The elements relationship to Dr Steiner's Agriculture Course**

C) **The elements relationship to the energetic body activities indicated by the Biodynamic Gyroscope**

D) **The polarity relationships arising from the circular periodic table**

Together, it is hoped, that these cross reference bases, can provide an image of the energetic activity of the chemical elements, based upon archetypal law, rather than randomised experimentation. These 'hints' of what these elements could be, can then investigated further. My investigations to date have shown that this is a worthwhile pursuit.

Jan Scholten

We also have available to us, the information developed by the homeopathic community, and particularly Jan Scholten. His book "Homeopathy and the Elements" (and others) are valuable references. Homeopathy has been 'proving' the effects of the chemical elements for some 200 years. Scholten's approach is based upon seeing the elements as a developing spiral. Starting with Hydrogen in the middle, and moving outwards with the atomic weight.

This provides seven layers within the spiral. He moves the Layers through a psychological developmental model from the baby through family , societal and then Humanitarian stages of development.

Within each layer, the elements moves through a process of expansion and then contraction, expressing the 'lesson' of that layer. I am in agreement with this basic model.

Homeopathy is a science that has developed along the tracks of observing symptoms from overdose circumstances, of a particular element. This is classified according to physical, emotional and psychological symptoms. They rarely speak of a substance as an energetic body interaction, so their excellent observations, can be translated further to see how, what they say can be described as an expression of an energetic activity.

I agree with his basic proposals of these sequences for the Major Elements and the Radioactive Elements, however I make a significant departure from Scholten, with the Transition Elements. I would like to leave how, to later, so as to not spoil the surprise. A second area of departure is that Glenopathy sees the Periodic Table as a 3D Torsian Sphere. Rather than a continuous spiral, the three groups of elements, make up one plane each within this 3D sphere. This gives each a definite 'inner journey'. Scholten merges the Transition Elements as part of the Major Elements experience, however with the Radioactive Elements, he gives them their own internal group experience.

The Power of Forms

Being one who works strongly with 'archetypal forms', I am aware of the possibility of holding firmly to the 'doctrine of the form', and thus deluding myself to the real experience of the element. Quantum Physics says they have proven, creation manifests according to the desires of the observer. So all 'subjective provers' and energetic investigators, need to hold back their will and desires, as best they can, so the element can speak its own voice. It's tricky and who am I to say how good I am at it. I know this is a very real phenomena.

This is a question I also have of Scholten, and his followers, too. They have a very strong commitment to 'the form', and one wonders how much they are willing the manifestations of the Elements they experience. A type of collective intention, that has them seeing down one direction, but not the other three.

My 'diversions' from their path, is not saying either of us is right or wrong, but that we have found different harmonics, of that particular frequency. Astrology is a science of different techniques, being applied to the same thing , an astronomical moment, and coming up with different parts of the same story. Often it is very similar parts of the same story. All of the information is most usually correct, within its context. My context is working with fundamental growth processes of Nature, whereas Homeopathy is most often Human psychology and physical symptom based.

The Layers

The Periodic Table has seven layers for each of its primary arms. The Biodynamic Vortex has six layers. However in my diagrams of the Vortex, the sixth layer of the zodiac, has two defined layers. In the Gyroscope diagram itself, this 12 fold zodiac layer is the purple ring at the very outside. As a Vortex comes to the top it curves in all directions, to then go onto form the spherical outer skin, of the gyroscopic sphere. This means that while the vertical distance of this purple area is the same as the other layers, the horizontal distance is extended. It is in this extended zone that this two fold division within the Spirit sphere is possible. This means that layers 6 and 7 of the Periodic table, are both encompassed in the 12 fold Spirit sphere of the Biodynamic Vortex.

The layers indicate a growing complexity, in the development of the chemical elements. This ccan be imaged as the development from the single celled organism to the complexities of the cosmos. The BD Vortex and Scholten have this complexity moving from the personal outwards to the global and cosmic. The elements themselves move from the very lightest, starting with Hydrogen to the very heavy.

In my pictures, there are 6 primary rings (p. 39), however each of these rings are dual in nature, which provides an image of the gyroscope with twelve layers. Reference to the Chartres Labyrinth with its 12 circles is helpful here (p. 212).

Layer 1 – The Core

H 1 • He 2

Layer one is represented in all my diagrams as the red layer, and in the Periodic Table it includes the elements of Hydrogen and Helium. I imagine this as Hydrogen in the middle, with Helium being on the second ring of the first main layer.

This is the primary core of all entities. If we are to look at life on Earth then it represents the Earth. For the Solar System, it is the Sun, for the atom the nucleus, for the Galaxy the Galactic centre.

Scholten describes this layer as "The Hydrogen Series" whose main concern is "to be or not to be". Should one incarnate or not? Other homeopathic texts describe Hydrogen, the most dominant element in space, as a very spiritual element and provides a universal connection or disconnection with all things. Scholten's progression through the layers has this corresponding to the conception and foetus phase of human development.

Dr Steiner's comments on Hydrogen has it as the carrier of the spirit impulses, both into and out of life. This means it is the carrier of the basic archetypal 'word' from the stars, upon which life is then built. If we see life forms as an expression of fractal magnetism – to use Dan Winters words – then Hydrogen is the carrier of the base formula of the fractal. In all life forms this is the fractal derived from the 'Golden Mean' ratio. Hydrogen carries the 'seed thought' of any particular individual.

Layer 2 – Duality, Earthly Substances

Li 3 • Be 4 • B 5 • C 6 • N 7 • O 8 • F 9 • Ne 10

This is the orange coloured layer of my diagrams, and is characterised in the Biodynamic Vortex, as the twofold layer of duality that stands outside of all life forms. Life manifests through the interplay of opposites. As soon as movement of any substance occurs, we see the development of polar opposite electrical charges, which then governs that particles interaction with all other substance. Cells pulsate and divide into two before dividing into four. Many animals only give birth to further physical bodies through the interaction of the two sexes of their species. This duality is not internalised in these cases and therefore remains an external reality existing 'prior' to life.

The elements of this layer; Lithium, Beryllium, Boron, Carbon, Nitrogen, Oxygen and Fluorine are recognised as some of the most basic in the structure of life, and most abundant in the Universe. Structurally these elements consist of the nucleus and one electron ring. This allows the charge of the nucleus to be strong and combine easily with most other elements. These elements are considered to easily give and receive electrons, so allowing for their highly reactive nature.

Along with Hydrogen, their fractal base, Carbon, Oxygen and Nitrogen are the basis of all organic chemistry. All Carbohydrate, Sugars, Fats and Proteins are formed primarily of these four elements. These four elements, Dr Steiner showed, are the basic carriers of the four bodies that come together to form living entities. Hydrogen carries the Spirit; Carbon is the basis of the physical bodies; Nitrogen carries the Astral body; while Oxygen is the carrier of the Etheric body in all its manifestations. Thus these are the **Earthly Substances** from which all else arises.

The elements Lithium, Beryllium and Boron are not made in a Star. They are made by a process called Cosmic Ray Spallation. Here cosmic rays split Oxygen and Nitrogen atoms, in space, in the atmosphere and in the soil, breaking these two elements down to these smaller elements. This phenomena indicates this ring of elements can be quite different to the other rings.

Scholten characterises this layer as the Carbon Series and suggests it is the stage of the individual becoming aware of themselves and the other. Who am I and where do I stand with the other, are the questions of this layer. It is the stage of the toddler and the early formation of the body.

Layer 3 – The Physical Body
Na 117 • Mg 127 • Al 137 • Si 147 • P 157 • S 167 • Cl 177 • Ar 18

The Yellow layer of the BD Vortex is the three fold layer, which manifests most strongly in the orientation and formation of the physical bodies of living organisms on our planet. This threefold process is the principle characterised as Thesis, Antithesis and Synthesis, in philosophic terms and manifesting as the primary duality between the Head, Nerve Sense system and the Metabolic Digestive system, finding their harmony in the Chest, breathing and circulatory rhythmic system. This middle zone is the result of the interplay between polarities at layer 2. Here we have the formation of physical bodies.

All the chemical elements in this Periodic Group are essential building blocks of the life forms we have around us, and each element can be said to work as a significant guiding principle in the various parts of our life system. Eg Silica carries the structural images other activities form around. Sodium and Phosphate are key elements in the nerve system. Magnesium and Sulphur are key elements of sea water and a healthy immune system. Hauschka says of Magnesium that it compresses life into solid earthly form. Of Sulphur we only have to reflect on its essential catalysing role in protein formation to see little would manifest without Sulphur. Aluminum, plays an essential role in harmonising the interplay of Calcium and Silica, those two essential elements and processes of all life forms (see Alchemical Chemistry).

Scholten characterises this as the Silicium Series and is an image of the teenager stage of human development. This is when the individual finds their relationship to their immediate community. Firstly their home and then their society of friends. This would be generally the age from 7 to 15 years.

Layer 4 – The Etheric Body

K 19 • Ca 20 • Ga 31 • Ge 32 • As 33 • Se 34 • Br 35 • Kr 36

Sc 2 • Ti 22 • V 23 • Cr 24 • Mn 25 • Fe 26 • Co 27 • Ni 28 • Cu 29 • Zn 30

Layer four is the layer of the Etheric body. This is where the fourfold law works most strongly, and allows for physical substance, to become life forms. At level four we have sustainable life. Here we have plant life coming into form. They are physical etheric beings with the Astral and Spirit activities still outside themselves. The etheric body is the basis of the immune glandular system and essential for good health.

The top layer of elements indicated here, are all elements that either strongly support or debilitate the etheric body. On the positive side, Potassium, Calcium, Germanium, Selenium are all common health promoters. Arsenic and Bromide in particular are well known for their poisonous effects; with both causing strong tiredness and exhaustion symptoms, an image of a weak etheric activity. In minute dosage, Arsenic is provided to chickens to stimulate weight gain.

The second line, are the first line of the transition or 'trace elements', which all act as specific catalysts in the processes of life. These are often called the 'Brothers of Iron' due to their relationship to this primary of all elements, for good blood formation and health. This is significant that the elements associated with stimulating specific life processes, come into manifestation, at his etheric level.

Scholten calls this the Ferrum series and identifies this period as the time of the young adult. From the years of 14 to 21 when an individual is learning a skill and becoming a productive member of ones society. In short, feeding life producing energy back to ones environment, responsibly. This is still at a village or small town level where all people in the environment know each other, and their place in society is determined by their skills and usefulness.

Layer 5 – The Astral Body
Rb37 • Sr38 • In49 • Sn50 • Sb51 • Te52 • I 53 • Xe54 • Y 39
Zr 40 • Nb41 • Mo42 • Tc43 • Ru44 • Rh45 • Pd46 • Ag47 • Cd48

The fifth layer is the level of the Astral body. It originates with the Solar system and has the planets Moon to Saturn as its 'organs'. This layer is where nitrogen enters deeply into action with Carbon , Hydrogen and Oxygen to develop proteins in the form of animal and human albumen. Here the animals are 'created', and the formation of organs becomes the indicator as to just how deeply the astrality has penetrated an organism. With the astrality comes consciousness and the psychic, psychological and sensation based realities.

The Astrality is not necessarily good for life forms. It acts as a stimulating and organising principle, however it does this by consuming the etheric activity. Very few of these elements would be considered beneficial for life, with only Tin, Silver and Iodine being considered to have some benefit, and then only in very small amounts.

Scholten calls this the Silver series and suggests its main function is in the passing on of ideas, be it through Art, advising, mysticism, music and "channeled information from other spheres". These are all activities of the stimulated senses and the last is a direct image of communication through the Astrality and its element nitrogen. Dr Steiner called Nitrogen and the Astrality 'a very clever fellow, it knows all things' due to the astrality's ability to be a dimension that knows no time or distance. Each part is in continual contact with all other parts simultaneously.

This is more middle age, where the individual is acting on a provincial level, characterised in an environment where we can no longer know everybody.

Layer 6 – The Collective Unconscious
Cs 55 • Ba 56 • Tl 81 • Pb 82 • Bi 83 • Po 84 • At 85 • Rn 86

La57 • Hf72 • Ta73 • W74 • Re75 • Os76 • Ir77 • Pt78 • Au79
Hg80 • Ce 58 • Pr 59 • No 60 • Pm 62 • Sm 63 • Eu 64 • Gd 65
Tb 66 • Dy 67 • Ho 68 • Er 69 • Tm 70 • Yb 70 • Lu 71

Between the Astrality and the Spiritlands there exists a zone Carl Jung called the Collective Unconscious. It has Uranus, Neptune and Pluto as its planetary masters and is a state where the collective oneness of the spiritlands, are merged with the Unconscious responsive aspects of the Astrality. It really is an in-between zone, where the individual goes on the spiritual journeys of the Occultist, Mystic and Shaman, in the process of bringing the spirits reality to full consciousness. My experience of these elements suggest they encourage the Internalised Spirit to incarnate.

All of these elements, except gold, are highly toxic or radioactive. The Rare Earths are also on this layer.

Scholten calls this the Gold series, where the urge is for power and leadership, and with this, comes responsibility. It is the area of the whole country and the period of old age.

Layer 7 – The Spiritlands

Fr 87 • Ra 88 • Ac 89 • Rf 104 • Db 105 • Sg 106 • Bh 107 • Hs 108 • Mt 109 • Ds 110 • Rg 111 • Cn 112 • Nf 113 • Fl 114 • Mc 115 • Lv 116 • Ts 117 • Og 118

Th 90 • Pa 91 • U 92 • Np 93 • Pu 94 • Am 95 • Cm 96 • Bk 97 • Cf 98 • Es 99 • Fm 100 • Md 101 • No 102 • Lr 103

This is the second stage of the sixth layer of the Biodynamic Vortex and indicates the Galaxy proper. All the elements are radioactive and highly toxic. Scholten calls this the stage of Magic. Here we have people who can reach their goals through the powers of thought and intention. Here one is fully conscious of the forces active in creation and one can work with them to bring about manifestations. I have linked the activities of this realm to the planets Persephone, Vulcan and the Sun in my "The Twelve Planets" chapters. This is very old age and the connections an individual has to the whole of humanity, and the wider Cosmos.

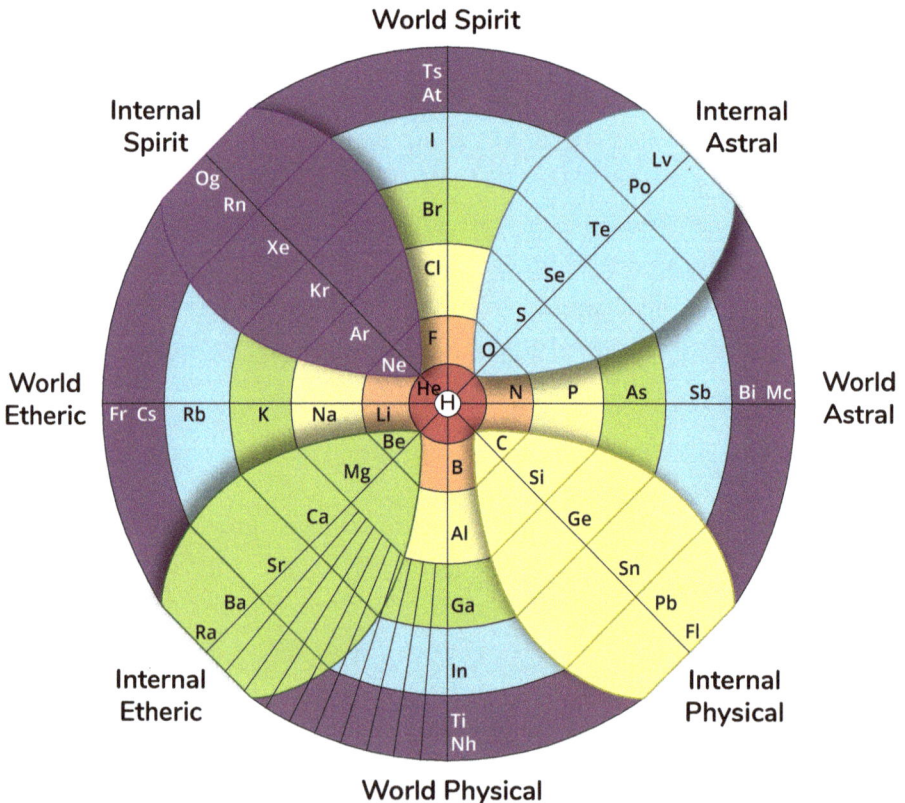

The Arms

The arms of the periodic table present an interesting development. With this development comes the possibility of a greater clarification, but also the possibility of a degree of confusion.

The reference system for the arms, arising from what has been outlined so far, has the arms of the Periodic table, ordered according to the internal and external activities of the energetic bodies. This seems to provide a very neat fit to the periodic table in most respects. This diagram (p. 78) indicates the spiritual activity of each of the Arms according to my previous reference systems. Dr Hauschka's 'Nature of Substance' (see 18, p. 238) has been used to clarify the associations in this diagram.

The Steiner Periodic encouraged by David Robison

The one element, which bucks the trend of a perfect fit with Dr Steiner's comments of the elemental carriers of energetic activity, and challenges this organisation, is the placement of oxygen.

In this diagram, Oxygen is found on the second layer of the Internalised Astral arm. This would suggest it is a significant element, that could be expected to help anchor the Astrality into physical forms. Therefore a question worth asking is **"How does Oxygen help anchor the astrality into life forms?"**

The significant dilemma this question poses, is caused by RS's very clear statement, (and I am in no doubt,) that Oxygen is the essential element for the carrying of the Etheric activity into life forms. In this context he showed that Carbon is the basis of the Physical forms, Nitrogen is the carrier of the Astrality, and Hydrogen are the basis of the Spirit's activity. A quick look at this diagram shows Nitrogen is in the right place and so is Carbon. Hydrogen is in the middle and so relates to all arms. A study of the Halogens will show, Hydrogen and the action of the Spirit on life-forms, have many similarities of action, to the very acidic and reactive Halogens.

So the one 'chink in the chain' of the Gyroscopic Chemistry presented so far is oxygen. I would prefer it was where Beryllium is, however being an anion, Oxygen has to be on the right hand side of the diagram, which has it directly opposite to Beryllium. Interesting! So far in my investigations into the elements this is the only element that does not fit the schema. So rather than declare the whole schema wrong, I ask, **"How is Oxygen right in this position?"**

A basic principle in Gyroscopic Astrology, when working with these archetypal forms, is **"It is All right"**. Because we are dealing with archetypal structures of universal proportions, there is probably far greater intelligence inherent in these models than my limited abilities. So I suggest we look for and trust this 'higher' intelligence, rather than declare our own understandings superior. In 'the game', wherever there is a piece that 'does not fit', in every previous instance to date, the challenge to find where it is right, has opened a door to another insight, or a whole new dimensional perspective, which positively adds unexpected depth and insight to the overall task. So it is with oxygen.

The Threefold law and the 3 Stages

In the Zodiacs section of "Biodynamics Decoded" I identified a three stage unwinding process, in regard to the zodiac diagrams. This arose from the fact that the ordering of the zodiac which arises out of the Astrological model, is different from the ordering of the zodiac constellations we find in the sky. If one starts with the Archetypal pattern of the planets and zodiac, a further two stages of unwinding processes provides the planets and zodiac order we have in the sky (see next page).

The splitting of a cell provides a similar image in life. The cell is 'content'. It then begins to move and pulsate, before splitting into two. This suggests there is an archetypal process of 3 stages, that can be found in various places.

Rudolf Steiner talks in 'Theosophy' of the 3 Energetic Worlds, within which we exist. I take him to be describing the Galaxy, The Solar System and the Earth as energetic dimensions active in the same Space and Time, but existing individually and interactively. The Life processes which give us the 4th Etheric World, comes from the Earth, moving outward. This 'idea' is an Asronomical FACT.

Stage 1 - Zodiac

Planets		Zodiacal Constellations				
♄	Saturn	♒	Aquarius		Capricorn	♑
♃	Jupiter	♐	Sagittarius		Pisces	♓
♂	Mars	♈	Aries		Scorpio	♏
☉	Sun					
♀	Venus	♎	Libra		Taurus	♉
☿	Mercury	♊	Gemini		Virgo	♍
☽	Moon	♌	Leo		Cancer	♋

Stage 1 - Zodiac

Stage 2 - Zodiac

Planets		Zodiacal Constellations				
♄	Saturn	♒	Aquarius		Capricorn	♑
♃	Jupiter	♓	Pisces		Sagittarius	♐
♂	Mars	♈	Aries		Scorpio	♏
☉	Sun					
♀	Venus	♎	Libra		Taurus	♉
☿	Mercury	♍	Virgo		Gemini	♊
☽	Moon	♌	Leo		Cancer	♋

Stage 2 - Zodiac

Stage 3 - Zodiac

Planets		Zodiacal Constellations				
♄	Saturn	♒	Aquarius		Capricorn	♑
♃	Jupiter	♓	Pisces		Sagittarius	♐
♂	Mars	♈	Aries		Scorpio	♏
☉	Sun					
♀	Venus	♉	Taurus		Libra	♎
☿	Mercury	♊	Gemini		Virgo	♍
☽	Moon	♋	Cancer		Leo	♌

Stage 3 - Zodiac

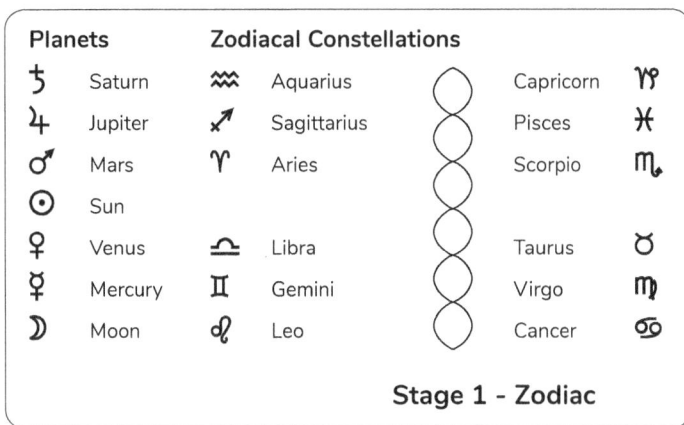

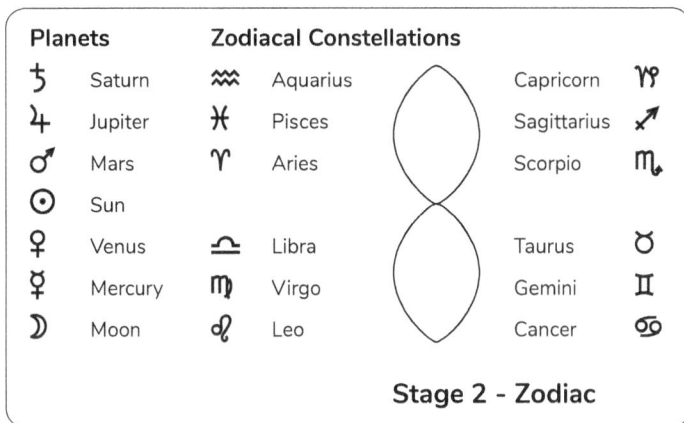

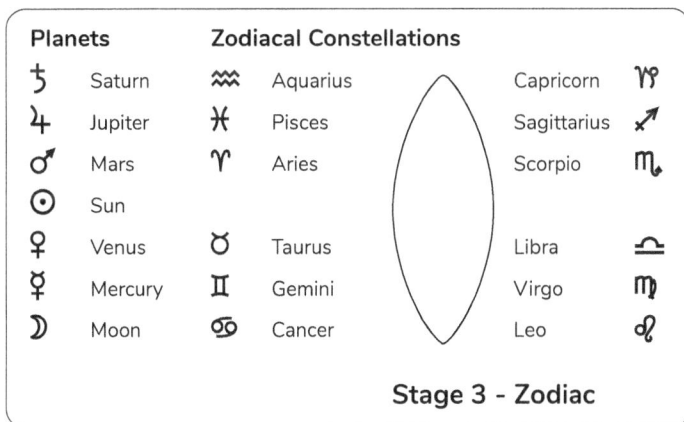

This three stage development was summarised as the first stage providing the archetypal law; then the in between 'lemniscate' stage, which stands behind manifestation, arises once there is movement. The third stage or the (en) unfolded stage, provides the form appropriate for manifestation. So far in our journey we have found **Stage 1, the Cosmic Rings phase and Stage 2, the Gyroscopic World phase**.

Where is oxygen RIGHT? I am in no doubt RS's statement that oxygen is the carrier of the etheric body; nitrogen the astral and so on, is correct. However we must note the context in which he said this. It was with regard to how the basis of matter forms. The substances Carbohydrate and Protein are made up primarily of these 4 substances. He is talking about manifestation. So lets follow what unfolds if we look at the Periodic Table anew, and add a third independent Manifesting Stage of activity. **What if these elements are the rulers of their arms**, which includes both the cation and anion sides of the axis.

Thus we have a Cation side and an Anion side of each energetic body arm. These could be expected to act as an inner polarity of each of these energetic body activities. The oxygen group including Sulphur and Selenium is the anionic Etheric arm, to the Magnesium and Calcium group, which is the cation Etheric arm. Sulphur and Selenium are both considered detoxifiers, which is what the Etheric body does.

This same process can be seen with the Astral arm. Nitrogen and Phosphorus are obvious elements of the astrality, while the Sodium, Potassium arm, with their central role in the working of the nerves, flowering and fruiting of plants, indicates they too have an obvious astral activity, inherent in their character.

This therefore suggests there are two major reference systems available for the Arms. **The spinning Gyro and this Manifest layer.** This double activity can represent two separate dimensions active 'in the same place'. One dimensional layer (the manifest) is most likely a result of the activity of the other.

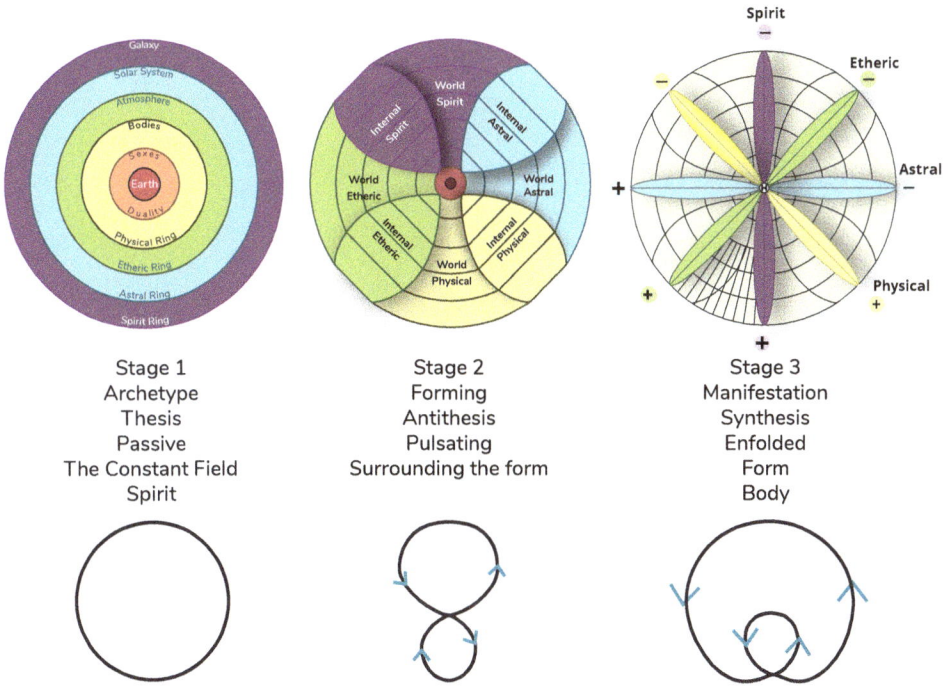

Stage 1	Stage 2	Stage 3
Archetype	Forming	Manifestation
Thesis	Antithesis	Synthesis
Passive	Pulsating	Enfolded
The Constant Field	Surrounding the form	Form
Spirit		Body

The Gyroscopic arms have come from cosmic references and associations, while the Steiner Arms have come from observing the basic manifest elements of the proteineous living activities in our environment.

By joining these two reference systems together we are provided with a 'World' energetic body activity pulsating at the base of these 'Environmental' arms activities.

The Steiner Etheric arm's 'environment' has a substrata supporting it, based on the interaction between the internal Etheric and Astral bodies (remembering the Astral motivates the Etheric).

The Astral arm has an internal substrata of the External Etheric and Astral.

The Spirit arm has a substrata of the External Spirit and External Physical activities.

83

The Physical Arm has a substrata of the Internal Physical and Internal Spirit activities.

These two reference systems are in addition to the first 'layer' of the External Cosmic Rings. **All three Stages can be used in the process of identifying the elements activities.**

Sulphur for example, would be considered a Physical anchor of an Astral activity within an Etheric environment. Selenium is a Etheric anchor of an Astral activity working within an Etheric environment.

Calcium is an Etheric / Etheric element in the Etheric environment. So we would expect this element to be an important element for the foundation of life. (Funny that) and so on.

Oxygen can be predicted to be a primary element that facilitates the Etheric bodies activity, while working within an Astral 'milieu'. The Astral being the active agent in the partnership. Oxygen becomes an element of 'Cosmic Substance' or primary anchor of an Astral activity within a Etheric environment.

We have to consider that the Etheric and Astrality, have a very intimate relationship in life forms. They ALWAYS work together in a push pull sort of relationship. Where one is strong the other is pushed out. Where one weakens the other moves in to fill the space left. In lifeforms, as in chemistry, oxygen acts as the fuel source for the other elements. Oxygen is the element that binds to all others until the element is 'oxidised' and becomes stable. Similarly, in the body, excessive nitrogen will consume the oxygen in an attempt to stablised. The Astrality and Spirit use the Etheric as fuel and once they have consumed too much of the etheric, the individual becomes exhausted. If it is totally consumed they die. Eg drug overdose.

Life does not come into form until the physical is taken up by etheric activity. This gives us the most basic form of single cell plants. Dr Eugen Kolisko (11) concluded that evolution through plants and through into the animal kingdom is an image of the degree of incarnation of the astral and ultimately the spirit into the lifeforms. So the astral does not come into the sphere of life unless there is an etheric body to receive it. It would appear this relationship between Nitrogen and Oxygen is the key to where the Astral and Etheric interaction occurs.

Sulphur, Oxygen's sister element is said by RS to be the 'oil' element in the C,O,N,H family. While these elements carry the spiritual bodies, it is Sulphur which acts as the lubricant that allows for them to work together. The role of Sulphur can be seen where a lack of Sulphur, leads to a seizing up of the bodies function, which leads to Autism. While too much Sulphur leads to a sloppiness in the bodies interaction leading to Hysteria. Mental / psychological illness can be seen as a manifestation between these two poles, and Sulphur is the key, 'to keep things moving'. We also know that Sulphur is the element in many biochemical reactions and the formation of protein and amino acids. It is the element that allows the astrality to become involved in the physical world. So we can anticipate that Oxygen, its relative, will have a similar role in smoothing the astrality's path into matter. It provides the etheric 'fuel' and doorway for it to enter the physical.

This is also interesting when considering RS comments in the third lecture regarding the close relationship between Nitrogen and Oxygen.

The difference between the External or Internalised body elements

In the Steiner Periodic schema, the Cardinal cross of the vertical and horizontal axis are the 'Cosmic' elements of the Spirit and Astrality, while the Secondary cross of the diagonal arms, are the arms of the 'Earthly' elements of the Physical and Etheric bodies. This reference may provide an avenue to identifying some of the difference in the nature of the Internal and External arms of the various bodies, at Stage 2. This is important when exploring what the difference in activity between Sulphur and Phosphorus would be. Both are Astral elements of the physical body ring. However Sulphur works more internally in biochemical reactions , while Phosphorus is more concerned with transportation of the other elements and energy generation for life forms and appears in only two basic forms, while Sulphur has many combinations.

The Astral Arm

Many of these elements, most notably K, Na, P, Li are all associated with the working of the nerve sense system in humans and in plants. P acts as the light energy transfer while K enhances the astralised flowering and fruit processes of plants.

Left Horizontal Vortex: World Etheric & Astral +
H · Li · Na · K · Rb · Cs

This is the arm of the World Etheric . This realm exists in the Atmosphere of the Earth and works with the world elements and ethers found in the atmosphere. The atmosphere is a special region, as its special oxygen content has been developed as a results of the life processes taking place on the Earth. Firstly in the oceans, due to the Blue Green Algae, and then from the plant life on Earth.

Right Horizontal Vortex: World Astral & Astral –
N · P · As · Sb · Bi

This is the area of the World Astrality. This exists in the region of the Solar system, particularly out as far as Saturn. This activity works particularly in the realm of light and nitrogen is its carrier. It works from above, onto all beings often in an impressive manner, forming from the outside. The World Astral is the fabric of the psychic realms which join all beings together in a collective 'sensation' and psychic realm.

The Etheric Arm

The elements Ca, Mg, O, S and Se are all considered basic elements of health and a strong immune system., which are the images of the etheric bodies functioning.

Left Green Petal: Internal Etheric & Etheric +
Be · Mg · Ca · Sr · Ba · Ra

This is the arm of the internalised Etheric body. This is the life providing body that works tirelessly keeping life forms alive and healthy. It is carried into life via oxygen and water in particular.

Top Right Blue Petal: Internal Astral & Etheric –
O · S · Se · Te · Po

This is the area of the internalised astral body. This is the body that lifts plants into animals and turns carbohydrates into proteins. This brings sensation and psychological influences to life forms.

The Spirit Arm

The Halogens naturally manifest the reactive and acidic aspects of the spirit, while the B and Al in particular can be seen to provide the archetypal direction elements in matter. Hugh Lovels gives an explanation of Boron as the 'base' element, needed to start the reactive change of elements in plant growth. Boron is needed for Silica to be come active to lift Calcium on its way. Al is the base element of clay, with the other 'directive' elements Silica and Phosphorus.

Bottom Vortex: External Physical & Spirit +
B · Al · Ga · In · Ti

This is the area of the World Physical. This is the physical substance of the Earth itself.

Top Vortex: External Spirit & Spirit –
Fl · Cl · Br · I · At

This is the World Spirit. This region is from the edge of the Solar system to the edge of the Galaxy and beyond. As this is the region of the stars, which generates the primary electromagnetic forces we are bombarded with, this is the basis of the formative principles upon which all other manifestation coalesces around.

The Physical Arm

The physical side of this arm is obvious, Silica forms the skeletal structure for the Calcium to lay down the mass of life forms. Tin and Lead are both noble metals and carry specific primary formative impulses. The inert gases pose a problem as they are considered to not interact with anything, so I will leave this as a question. What is the dynamic relationship between the inert gases and the cation physical elements?

Right Yellow Petal: Internal Physical & Physical +
C · Si · Ge · Sn · Pb

This is the region of the internalised physical body. This occurs when the etheric body picks up the World Physical substance and brings it into sustainable life forms. This is the clay that the 'sculptor – the other bodies –' uses. These elements carry the structural impulses other elements can work around.

Top Left Petal: Internal Spirit – Physical –
He · Ne · Ar · Kr · Xe · Rn

This is the sector of the internalised spirit. Dr Steiner calls this the Ego, or higher Ego and is the I aspect of the individual. This is ultimately our consciousness and the eternal spark which Lord Krishna says in the Bhagavad-Gita, can not ever be killed.

The Electronic Being

In my explorations of 'The Electric Universe' I have been challenged by the findings of others who talk of a non electro magnetic (EM) 'dimension', where the normal laws of EM do not apply. Things move at greater than the speed of light, particles 'communicate' instanteously across huge distances, and this activity is highly supportive of life processes. This is called 'The Ether' by some, Orgone by the Riecheans, and Scalar or Di- electric by alt science. One thing in common with most of these commentators is that this activity is 'the opposite of' or higher companion of EM, with EM 'dropping out' of this 'higher' activity.

In the case of scalar at least, it is identified as being generated at a 45 degree angle to the electric and magnetic perpendicular relationship, just as electro magnetism is also. I suggest that both electro magnetism and Di magnetic activity are created as 'overtones' of the spinning activity of the Magnetic and Electric gyroscope. Thus they are siblings.

Thus even though this activity does not conform to the normal laws of EM, it is not separate from the Electronic Being. It is a manifestation of the electronic spherical being, at the basis of all manifestation, like everything else. Change the activity of the electric and magnetic poles and the Di Magnetic / scalar effect will also change.

I suggest that EM does not drop out of the Ether, as much as both **EM and Di-Electric** are manifestations of the Electric and Magnetic primary axis, movement. They 'anchor' themselves on the **Second Vertical Axis of the Gyroscope**, based upon the Zenith and the Nadir.

Putting the various images together suggests the top diagram, which has to be extended to a 3D image to gain a true image of the inner life of the Electronic Being.

All is One

Ts
At
Spiritual Formative Forces
World Spirit

Astral Lv
Formative Forces
Internal Spirit in Internal Astral
♂2 & ♃2
Po
Te
World Astral in World Spirit
♃1 & ♄1
I
Free Fire
World Etheric in World Spirit
Br
Cos. Si.
Phys. World in Spirit
Cl
Internal Astral
Se
Bound Light
S Internal Phys in Astral
Internal Etheric in Internal Astral
Cosmic Matter
F H O
Clay
Sand
Internal

Og
Rn
Higher Ego
Internal Spirit in Internal Spirit
Internal Astral in Internal Spirit
Xe
♄2 & ♃2
Kr
Bound Warmth
Internal Etheric in Internal Spirit
Ar
Ne
Int. Physis in Int. Spirit
Cosmic Int. Forces in Int. Spirit

W. Spirit in W. Etheric
W. Astral in W. Etheric
World Etheric
W. Phys. in in W. Etheric
O He H N N
Ter. Si.
P
Free Light
As
♂1 & ♃1
Sb
World Astral
Bi Mc

Fr Cs
Rb
♀1 & ☿1
K
Free Water
Na
Cos. Ca.
Li
Be
C B
World Phys in W. Astral
World Etheric in World Astral
World Astral
World Spirit in World Astral

World Etheric

Int. Phys. in Int. Etheric Forces
Mg
Earthly
Humus
C
Al
Lime
Si
Earthly Matter
Ge
Bound Life Ether

Ca
Internal Etheric
Bound Chemical Ether
World Phys.
Ter. Ca.
Internal Physical
Int. Etheric in Int. Physical
Sn
☽2 & ☿2

Sr
Int. Astral in Int. Etheric
♀2 & ☿2
World Etheric in World Physical
Ga
Free Earth
Internal Astral in Internal Physical
Pb
Physical Formative Forces

Ba
Int. Spirit in Int. Etheric
In
World Astral in World Physical
☽1 & ☿1
Internal Spirit in World Physical
Fl

Ra
Etheric Formative Forces
World Spirit in World Physical
Ti
Nh
World Physical

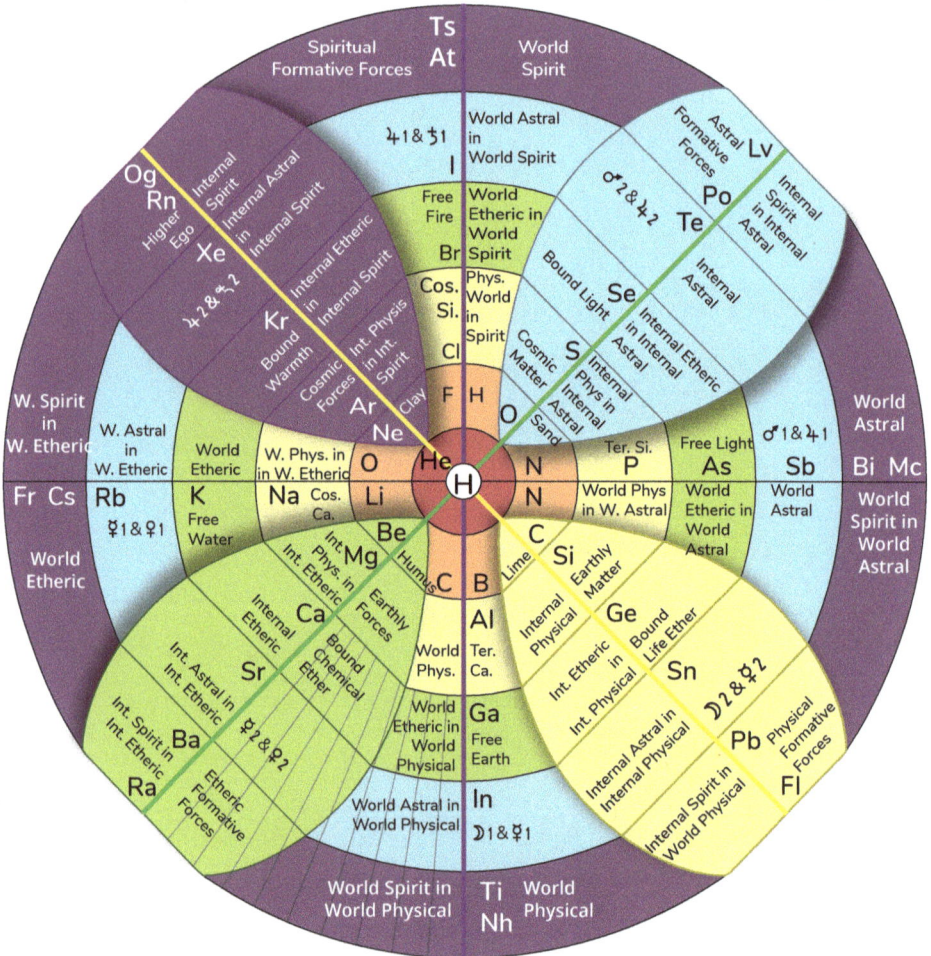

Agriculture Course

Spiritual Bodies Activity

Spirit
Astral
Etheric
Physical

Hydrogen H

Chemical Data: 1 • 1.0079
Steiner Arm: Spirit
Gyro Arm: Spirit
Layer: One
Ag. Course Name: Carrier of the spiritual
archetype
into the physical
Spiritual Activity: Spiritual Archetype

Helium He

Chemical Data: 2 • 4.0026
Steiner Arm: Physical –
Gyro Arm: Inert Gases –
Internalised Spirit
Layer: One
Homeopathy: Autism, Stay in themselves
and experience their own being. No
worldly values, they simply live.
Do not react to anything external.
Spiritual Activity: Initial point of
incarnation of the Spirit

Lithium Li

Chemical Data: 3 • 6.941
Steiner Arm: Astral +
Gyro Arm: External Etheric
Layer: Two – Cosmic Substance
Ag. Course Name: 'Oxygen' the anchour
of the Etheric to the
physical realm
Spiritual Activity: Anchor of the
World Etheric

BerylliumBe

Chemical Data: 4 • 9.0122
Steiner Arm: Etheric +
Gyro Arm: Internalised Etheric
Layer: Two – Cosmic Substance
Ag. Course Name: Humus
Spiritual Activity: Anchor of the
internalised etheric

Boron B

Chemical Data: 5 • 10.811
Steiner Arm: Spirit –
Gyro Arm: External Physical
Layer: Two – Cosmic Substance
Ag. Course Name: Carbon, the physical
basis of life forms
Spiritual Activity: Anchor of the
World Physical

Carbon C

Chemical Data: 6 • 12.011
Steiner Arm: Physical +
Gyro Arm: Internalised Physical body
Layer: Two – Cosmic Substance
Ag. Course Name: Lime, the anchoring
of the life to the physical body
Spiritual Activity: Anchor of the
Internal Physical processes

Nitrogen N

Chemical Data: 7 • 14.007
Steiner Arm: Astral –
Gyro Arm: External Astral
Layer: Two – Cosmic Substance
Ag. Course Name: Nitrogen, the carrier
of the astrality into matter
Spiritual Activity: Anchor of the
World Astral activity

Oxygen O

Chemical Data: 8 • 15.999
Steiner Arm: Etheric –
Gyro Arm: Internalised Astral
Layer: Two – Cosmic Substance
Ag. Course Name: Sand
Spiritual Activity: Stimulates the
Manifest Etheric activity

Fluorine F

Chemical Data: 9 • 18.998
Steiner Arm: Spirit –
Gyro Arm: External Spirit
Layer: Two – Cosmic Substance
Ag. Course Name: Hydrogen
Spiritual Activity: Anchor of the World Spirit into matter

Neon Ne

Chemical Data: 10 • 20.180
Steiner Arm: Physical –
Gyro Arm: Internalised Spirit
Layer: Two – Cosmic Substance
Ag. Course Name: Clay
Spiritual Activity: Anchor of the Internalised Spirit

Sodium Na

Chemical Data: 11 • 22.990
Steiner Arm: Astral +
Gyro Arm: World Etheric
Layer: 3 – Physical Body
Ag. Course Name: Cosmic Forces
Spiritual Activity: World Physical forces work upon the External Etheric

Magnesium Mg

Chemical Data: 12 • 24.305
Steiner Arm: Etheric +
Gyro Arm: Internalised Etheric
Layer: 3 Physical body
Ag. Course Name: Cosmic Substance
Spiritual Activity: Combines the Etheric into the Physical body

Alumina Al

Chemical Data: 13 • 26.982
Steiner Arm: Spirit +
Gyro Arm: External Physical
Layer: 3 Physical bodies
Ag. Course Name: Terrestrial Calcium
Spiritual Activity: Cosmic Physical into the World Physical activities

Silica Si

Chemical Data: 14 • 28.086
Steiner Arm: Physical +
Gyro Arm: Internalised Physical body
Layer: 3 Physical bodies
Ag. Course Name: Earthly Matter
Spiritual Activity: Cosmic Spirit into the Internal Physical activities

Phosphorus P

Chemical Data: 15 • 30.974
Steiner Arm: Astral –
Gyro Arm: External Astrality
Layer: 3 Physical bodies
Ag. Course Name: Terrestrial Silica
Spiritual Activity: Cosmic Physical into the World Astrality

Sulphur S

Chemical Data: 16 • 32.066
Steiner Arm: Etheric –
Gyro Arm: Internalised Astral body
Layer: 3 Physical bodies
Ag. Course Name: Cosmic Matter
Spiritual Activity: Cosmic Physical into the Internal Astral body

Chlorine Cl

Chemical Data: 17 • 35.453
Steiner Arm: Spirit –
Gyro Arm: World Spirit
Layer: 3 Physical bodies
Ag. Course Name: Cosmic Silica
Spiritual Activity: Cosmic Physical into the World Spirit

Argon Ar

Chemical Data: 18 • 39.948
Steiner Arm: Physical –
Gyro Arm: Internalised Spirit
Layer: 3 Physical bodies
Ag. Course Name: Cosmic Forces
Spiritual Activity: Cosmic Physical's influence upon the Internal Spirit

Potassium K

Chemical Data: 19 • 39.098
Steiner Arm: Astral +
Gyro Arm: World Etherics
Layer: 4 Etheric body
Ag. Course Name: Free Water /
Chemical ether
Spiritual Activity: Cosmic Etheric stimulates the World Etheric Forces

Gallium Ga

Chemical Data: 31 • 69.723
Steiner Arm: Spirit +
Gyro Arm: External Physical
Layer: 4 Etheric body
Ag. Course Name: Free Earth/Life ether
Spiritual Activity: Cosmic Etheric onto the World Physical

Arsenic As

Chemical Data: 33 • 74.922
Steiner Arm: Astral –
Gyro Arm: World Astral body
Layer: 4 Etheric body
Ag. Course Name: Free Air/Light Ether
Spiritual Activity: Cosmic Etheric activation of the World Astrality

Bromine Br

Chemical Data: 35 • 79.904
Steiner Arm: Spirit –
Gyro Arm: External Spirit
Layer: 4 Etheric Body
Ag. Course Name: Free Fire/Warmth Ether
Spiritual Activity: Cosmic Etheric working in the World Spirit

Calcium Ca

Chemical Data: 20 • 40.078
Steiner Arm: Etheric +
Gyro Arm: Internalised Etheric body
Layer: 4 Etheric body
Ag. Course Name: Bound
Chemical ether
Spiritual Activity: Cosmic Etheric stimulates the Internalised Etheric forces

Germanium Ge

Chemical Data: 32 • 72.61
Steiner Arm: Physical +
Gyro Arm: Internalised Physical Body
Layer: 4 Etheric body
Ag. Course Name: Bound Life Ether
Spiritual Activity: Cosmic Etheric stimulation of the Internal Physical body

Selenium Se

Chemical Data: 34 • 78.96
Steiner Arm: Etheric –
Gyro Arm: Internalised Astral
Layer: 4 Etheric Body
Ag. Course Name: Bound Light Ether
Spiritual Activity: Cosmic Etheric stimulates the Internal Astrality

Krypton Kr

Chemical Data: 36 • 83.80
Steiner Arm: Physical –
Gyro Arm: Internalised Spirit
Layer: 4 Etheric Body
Ag. Course Name: Bound Warmth Ether
Spiritual Activity: Cosmic Etheric working into the internalised Spirit

Rubidium Rb

Chemical Data: 37 • 85.468
Steiner Arm: Astral +
Gyro Arm: World Etheric
Layer: 5 Astral Body
Ag. Course Name: Mercury 1 & Venus 1
Spiritual Activity: Cosmic Astrality stimulates the World Etheric

Strontium Sr

Chemical Data: 38 • 87.62
Steiner Arm: Etheric +
Gyro Arm: Internalised Etheric
Layer: 5 Astral Body
Ag. Course Name: Mercury 2 & Venus 2
Spiritual Activity: Cosmic Astral stimulates the World Etheric activities

Indium In

Chemical Data: 49 • 114.82
Steiner Arm: Spirit +
Gyro Arm: External Physical Body
Layer: 5 Astral Body
Ag. Course Name: Moon 1 & Mercury 1
Spiritual Activity: Cosmic Astrality stimulates the World Physical

Tin Sn

Chemical Data: 50 • 118.71
Steiner Arm: Physical +
Gyro Arm: Internalised Physical
Layer: 5 Astral Body
Ag. Course Name: Moon 2 & Mercury 2
Spiritual Activity: Cosmic Astral body stimulates the Internal Physical Body

Antimony Sb

Chemical Data: 51 • 121.76
Steiner Arm: Astral –
Gyro Arm: World Astrality
Layer: 5 Astral Body
Ag. Course Name: Mars 1 & Jupiter 1
Spiritual Activity: Cosmic Astral stimulates the World Astral activities

Tellurium Te

Chemical Data: 52 • 127.60
Steiner Arm: Etheric –
Gyro Arm: Internalised Astral
Layer: 5 Astral Body
Ag. Course Name: Mars 2 & Jupiter 2
Spiritual Activity: Cosmic Astral stimulates the Internal Astrality

Iodine I

Chemical Data: 53 • 126.90
Steiner Arm: Spirit –
Gyro Arm: World Spirit
Layer: 5 Astral Body
Ag. Course Name: Jupiter 1 & Saturn 1
Spiritual Activity: Cosmic Astrality stimulates the World Spirit

Xenon Xe

Chemical Data: 54 • 131.29
Steiner Arm: Physical –
Gyro Arm: Internalised Spirit
Layer: 5 Astral body
Ag. Course Name: Jupiter 2 & Saturn 2
Spiritual Activity: Cosmic Astral stimulates the Internal Spirit

Cesium Ce

Chemical Data: 55 • 132.91
Steiner Arm: Astral +
Gyro Arm: World Etheric
Layer: 6 Internal Spirit
Ag. Course Name: Aries
Spiritual Activity: Internal Spirit directs the World Etheric activities

Barium Ba

Chemical Data: 56 • 137.33
Steiner Arm: Etheric +
Gyro Arm: Internal Etheric
Layer: 6 Internal Spirit
Ag. Course Name: Libra / Gemini
Spiritual Activity: Internal Spirit directs the Internal Etheric

Thallium Tl

Chemical Data: 81 • 204.38
Steiner Arm: Spirit +
Gyro Arm: World Physical
Layer: 6 Internal Spirit
Ag. Course Name: Cancer / Leo
Spiritual Activity: Internal Spirit directs the World Physical

Bismuth Bi

Chemical Data: 83 • 208.98
Steiner Arm: Astral –
Gyro Arm: World Astral
Layer: 6 Internal Spirit
Ag. Course Name: Scorpio
Spiritual Activity: Internal Spirit interacting with the World Astral

Astatine At

Chemical Data: 85 • 209.99
Steiner Arm: Spirit –
Gyro Arm: World Spirit
Layer: 6 Spirit
Ag. Course Name: Aquarius / Capricorn
Spiritual Activity: Internal Spirit interacts with World Spirit

Francium Fr

Chemical Data: 87 • 223.02
Steiner Arm: Astral +
Gyro Arm: World Etheric
Layer: 7 Cosmic Spirit
Ag. Course Name: Aries
Spiritual Activity: Cosmic Spirit into the World Etheric

Nihonium Nf

Chemical Data: 113 • 286
Steiner Arm: Spirit +
Gyro Arm: World Physical
Layer: 7 Cosmic Spirit
Ag. Course Name: Cancer
Spiritual Activity: Cosmic Spirit into World Physical

Lead Pb

Chemical Data: 82 • 207.20
Steiner Arm: Physical +
Gyro Arm: internal Physical
Layer: 6 Internal Spirit
Ag. Course Name: Virgo
Spiritual Activity: Internal Spirit in the Internal Physical

Polonium Po

Chemical Data: 84 • 208.98
Steiner Arm: Etheric –
Gyro Arm: Internal Astral
Layer: 6 Internal Spirit
Ag. Course Name: Pisces
Spiritual Activity: Internal Spirit directs the Internal Astral body

Radon Rn

Chemical Data: 86 • 222.02
Steiner Arm: Physical –
Gyro Arm: 6 Spirit
Layer: Internal Spirit
Ag. Course Name: Sagittarius
Spiritual Activity: The Internal Spirit – I

Radium Ra

Chemical Data: 88 • 226.03
Steiner Arm: Etheric +
Gyro Arm: Internalised Etheric
Layer: 7 Cosmic Spirit
Ag. Course Name: Libra
Spiritual Activity: Cosmic Spirit in the Internalised Etheric

Flerovium Fl

Chemical Data: 114 • 289
Steiner Arm: Physical +
Gyro Arm: Internalised Physical
Layer: 7 Cosmic Spirit
Ag. Course Name: Virgo
Spiritual Activity: Cosmic Spirit into the Internal Physical body

Moscovium Mc

Chemical Data: 115 • 289
Steiner Arm: Astral –
Gyro Arm: World Astral
Layer: 7 Cosmic Spirit
Ag. Course Name: Libra
Spiritual Activity: Cosmic Spirit into World Astral

Livermorium Lv

Chemical Data: 116 • 290
Steiner Arm: Etheric –
Gyro Arm: Internalised Astral Body
Layer: 7 Cosmic Spirit
Ag. Course Name: Sagittarius
Spiritual Activity: Cosmic Spirit into the Internal Astral Body

Tennessine Ts

Chemical Data: 117 • 294
Steiner Arm: Spirit –
Gyro Arm: World Spirit
Layer: 7 Cosmic Spirit
Ag. Course Name: Capricorn
Spiritual Activity: Cosmic Spirit into World Spirit 'God'

Oganesson Og

Chemical Data: 118 • 294
Steiner Arm: Physical –
Gyro Arm: Internalised Spirit
Layer: 7 Cosmic Spirit
Ag. Course Name: Pisces
Spiritual Activity: Cosmic Spirit into Internalised Spirit

The Transition Elements

The 5th Harmonic – Life's toolbox

The transition elements pose an interesting question when placed on the gyroscope. Given the 8 primary arms are 'dominant', and are used to form the axis of the gyroscope, it means the transition elements fall as a group in the left hand bottom quadrant. This places them over the edge of the Internal Etheric and World Physical spheres (see p. 78) where life occurs.

When one looks at the activity of the transition elements, they are involved in acting as catalysts, for life processes in biochemistry. It is therefore no mistake for them to be placed where the 'life' processes interact with the 'dead' world physical processes. On top of this, their activation of the life sphere, corresponds with them coming into manifestation, at Layer 4, the Etheric / Life ring of the process.

These are the elements that facilitate the activity of life, and life likes lemniscates.

Dr Hauschka describes a very interesting phenomena. The Periodic Table develops according to increasing Atomic Weight. The hardness and the melting points of the elements follow along in syc with this phenomena. Until we come to the Transition Elements. Here these qualities do not follow the atomic mass increase of the elements. By melting point and hardness, it goes Calcium, then Zinc through to Scandium, then back to Gallium.

This ordering resolves one of the 'odd' phenomena found in the traditional periodic table. This phenomena is not addressed by most chemical commentators.

We saw earlier how life processes manifest through the movement of a lemniscate. It would therefore be reasonable to suggest our understanding of these 'elements of life' could be **expanded** upon by a lemniscate. This would provide a primary 'circle' of the major elements and a secondary circle of the transition elements, by **pulling them out and twisting them into a lemniscate**.

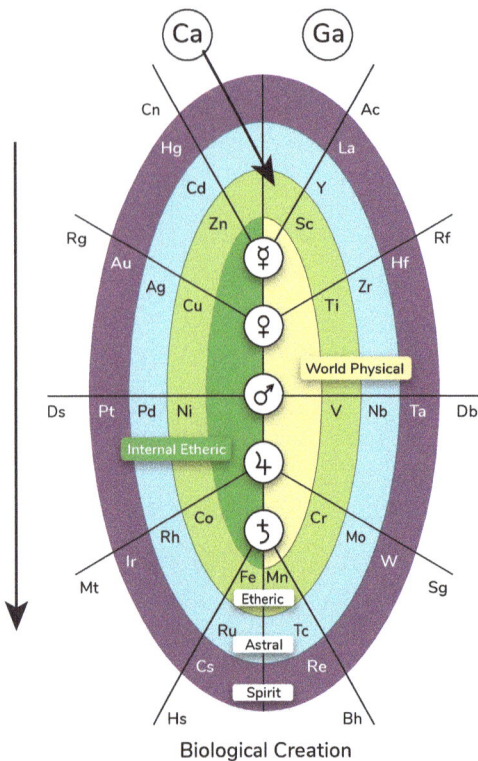

Biological Creation

This action achieves a couple of things. Firstly, it allows these elements to be **placed upon a circle**, which then brings them into the **same context as the gyroscopic** references. Secondly, due to the 'flip' that occurs in the lemniscating process, the **elements become reversed within the references to the Internal Etheric and World Physical arms**. Influencing how we might see them regarding the Incarnating and Excarnating references from Lievegoed.

In this picture, the flow of elements begins with Calcium (20) and moves to Sc (21) and around the circle till Zn (30) and on to Gallium (31).

This reverses the relationship of the trace elements to the 'major' gyroscope. The elements from 21 to 25 were previously on the Internalised Etheric side. They are now placed on the World Physical side of the diagram.

In keeping with RS indication for interpreting the practical use of these lemniscate flips, ala Twelve Sense – the elements can be read from Calcium to Zinc, Copper and around to Scandium before moving onto Gallium and onwards.

This suggests that due to their placement with the Internal Etheric activities we would expect the elements Zinc, Copper, Nickel, Cobalt and Iron, will be more active in the support of life processes, than those from Manganese, Chromium through to Scandium. The later elements would be expected to be more related to building structure and fixing forms. Hauschka identifies Mn, Cr and Va as elements that enhance sclerotic and hardening processes.

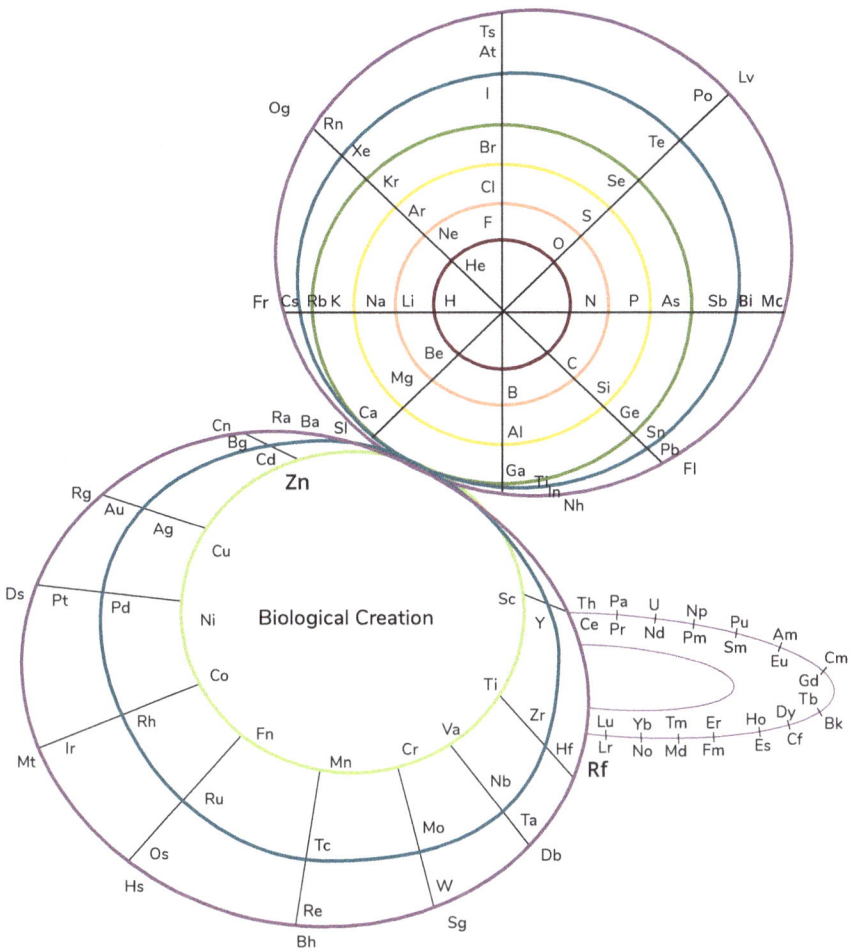

Ten Elements – Fifth Harmonic

There are **four bands/rings of transition elements, which provides a relationship to the overall energetic bodies via the RING they are placed in**. The first transition ring is the Cosmic Etheric ring, The second is the Cosmic Astral ring, the next is the first zodiac ring which is influential upon the Internal Spirit, while the fourth ring is the World Spirit again. In each of these ring sets there are ten elements.

The MAJOR gyroscope naturally falls upon the 8 arms. If we use music as a reference then the **MAJOR elements are in the second harmonic** of the circle, with its multiples of 2,4,8 being present in their relationships. The basic degree division on this harmonic of the circle is 45 degrees. The **trace**

elements on the other hand, being ten in division, suggests they are resonating to the fifth harmonic of the circle. The basic degree division of the firth **harmonic** of a circle is 36 & 72 degrees. The form arising from this five / ten fold division of a circle is the pentagram.

A range of five fold relationships can therefore be investigated to help identify their nature.

"Amongst the Hebrews, the five point symbol was ascribed to Truth and to the five books of the Pentateuch. The ancient Greeks, called it the Pent alpha. Pythagoreans considered it an emblem of perfection or the symbol of the human being.

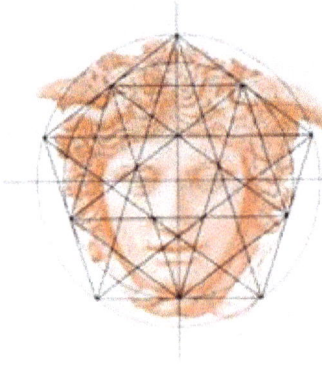

The pentagram is associated with the golden ratio (which it includes). The dodecahedron, the fifth Platonic solid, has twelve pentagonal faces, and was considered by Plato to be a symbol of the heavens."

This number can be expected to be an important indicator of life processes, easily seen in the five fold nature of the human body. Pythagoras, the father of mathematics, identified the relationship of the pentagram to the activities of life, through its exact expression of the 'Golden Mean' ratio. – **1 : 1.618** – It is well recorded that the 'Golden Mean' manifests in the relationships found in life forms and can be said to be the algorithm upon which our holographic creation manifests. This is seen in such things as the ratio of the length of the bones in the hand to each other, the ratio of the face and our view of beauty, the proportions found in plant growth, the share market movement and so on.

The Fivefold Venus Path

This pentagram picture, shows a tenfold division is represented as two pentagrams, one inside the other, which provides an image of the ten fold division we find in the trace elements. This is one indication that the ten elements could be split into two groups; Pentagram A and Pentagram B (a).

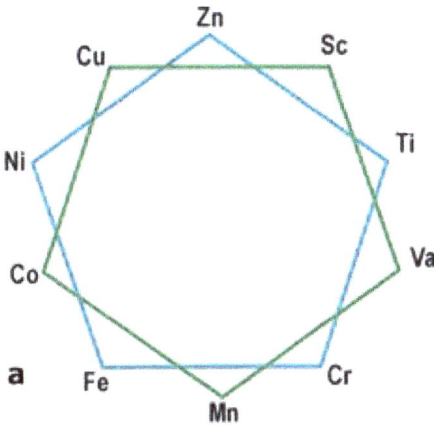

This division is given relevance by referencing the Venus related pentagrams, formed **by the Superior and Inferior conjunctions of Venus and the Sun** (b). Over an eight year period the path of these conjunctions forms two pentagrams in the sky. The diagram shows the Venus path. The obvious pentagram in the middle is the Inferior conjunctions while the Superior conjunctions occur when Venus is at the extremity of its cycle, so that pentagram is drawn on the outer circle.

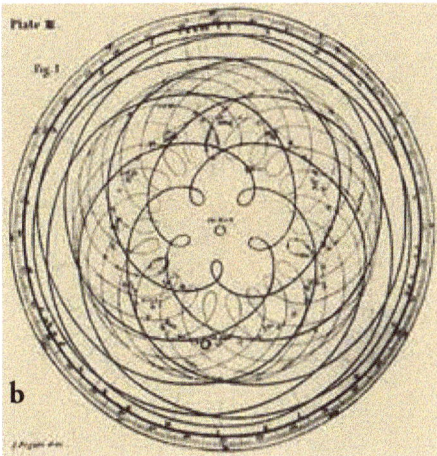

This provides an image of two separate pentagrams. In 'the' literature these pentagrams are often described as one pentagram with a single point to above, and the other with two points above. Pythagoras and others identified the five fold process as being made up of the four elements of Earth Air Water and Fire with Spirit being the single point above. This represents the four elements dominated or organised by spirit. Some occult groups have given this symbol to people who had achieved a high degree of initiation. The pentagram with the two points at the top is understood to represent matter dominating spirit, and was given to the lower level of initiates, who were still at the stage of their astrality dominating their spirit.

The Planets

Whenever a five fold series is identified it is possible to relate them **to the five planets from Mercury to Saturn**. In this regard the Sun and the Moon are treated as 'parents', while the other planets are their 'children'. The Moon could be allotted to Calcium and Gallium to the Sun, in this instance while

101

the other planets are assigned to two elements each (p. 96). This places the Moon with Calcium and a ruler of the Internal Etheric processes, while the Sun is placed with Gallium and rules the World Physical process of manifestation.

With this planetary reference comes the possibility of finding a relationship to the plant. This form of double activity of the planets, shown here, can be found in relation to plant growth in my chapters on 'Biodynamic Plant Growth', and was put forward in Biodynamic literature by Geothe, and developed further by Dr Lievegoed.

The primary process is seen as a developmental (Being) phase, which culminates in germination, followed by a secondary manifestation phase, where the form of the plant we are familiar with grows and finally reaches seeding.

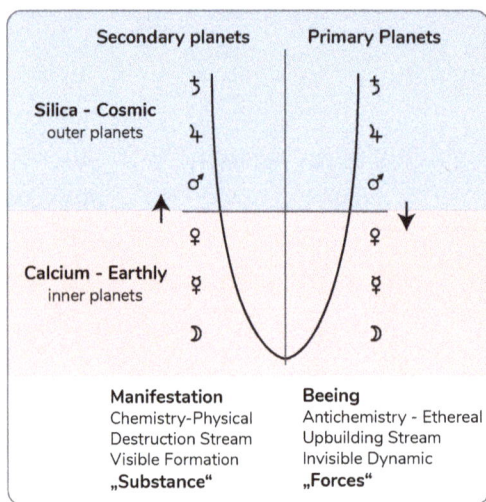

With the transition elements, this 'being' phase, is shown in the activities of the elements of the Internalised Etheric, while the 'manifest' phase is shown in the World Physical side of the group.

We can use the planets to deepen our understanding of the quality of the element and then possibly to the plant processes it might effect. This suggests that Zinc will be a primary Mercury activity, while Scandium will carry a quality of the secondary Mercury activity. In plants Mercury has to do with sap flow and expansive, running growth, Venus flowering, Mars, fertilisation and protein formation, Jupiter oil formation, while Saturn has to do with seed formation.

Reflecting back to the 2 pentagrams, this reference for the transition elements, having a coming into Being and a Manifest stages, can be reflected back to the 2 pentagrams previously identified. Which is which, I expect, will arise out of a closer inspection of the elements, however a suggestion can be made. The inferior conjunctions are those where Venus is between the Sun and the Earth, while the Superior conjunctions have Venus on the opposite side of the Sun to the Earth. So one has Venus' activity strong and close, while the other has Venus away, even blocked by the Sun. With Venus being a feminine planet, you would be expected to find less feminine and more masculine qualities associated with the Superior conjunction set. The Iron pentagram is the 'coming into being' side, and its group would be expected to be associated with the inferior conjunctions, and the blue pentagram – lets call them the feminine Pentagram A, while the Manganese set would become associated with the superior 'male' conjunctions – Pentagram B.

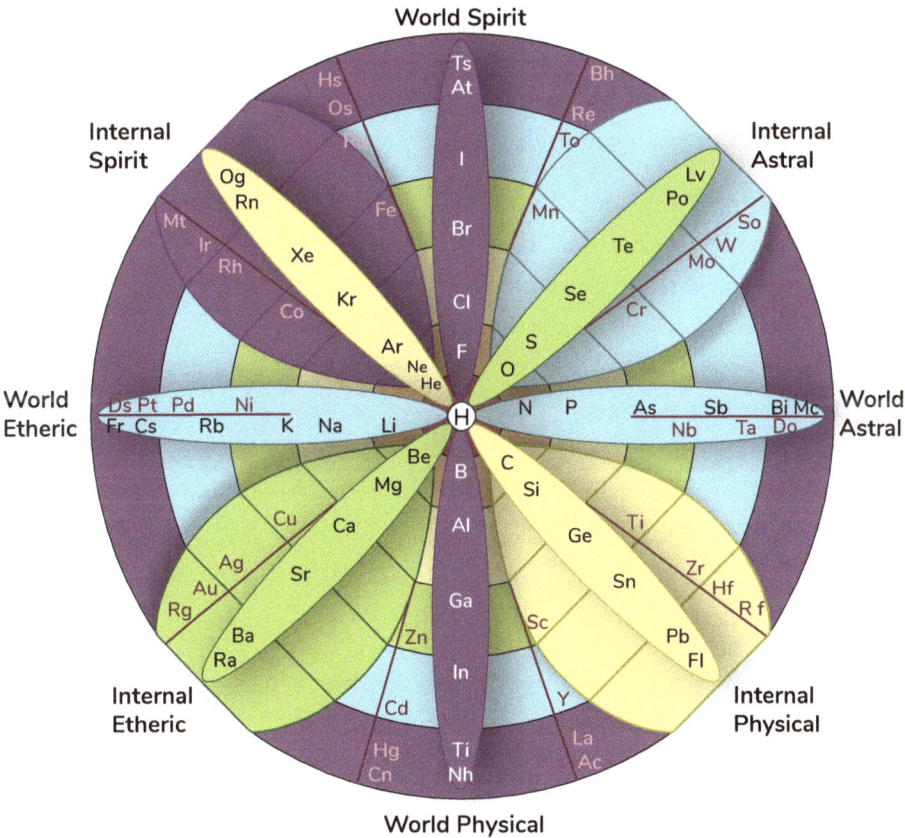

Polarities

Another relationship that would naturally be suggested for investigation from the fivefold division is the polarity or opposite relationships between the elements. How are these relationships significant?

Zinc – Manganese
Copper – Chromium
Nickel – Vanadium
Cobalt – Titanium
Iron – Scandium

Transition Elements and the Gyroscope

The life enfolding process comes in three stages. The first stage is a pushing out, the second stage is the flip to form the lemniscate, while the third stage is a (en) folding back upon itself.

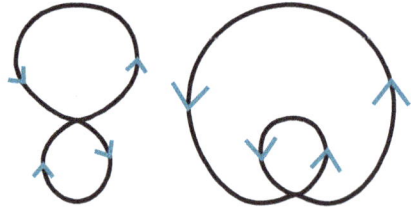

I have expressed this previously with this diagram. It is therefore reasonable to consider that the transition elements lemniscate, already discussed, can be folded back over the 8 fold gyroscope, thus providing a further cross reference. This provides a reference of the **transition elements to the energetic body activities**. So while there is a 'primary' association of these elements, as either Internal Etheric or World Physical dominated elements, there is now a secondary energetic body activity, indicating how these elements work upon the other bodies as well.

One more observation to make from this diagram is that **the transition elements are now placed close to major elements**, which we can assume is of the same spiritual body activity. An initial stand out is; Cu sitting next to Calcium; with Silver and Gold, now also being identified as elements which will stimulate the activity of the internalised Etheric body. This activity of these elements are all well known and commonly used. This diagram is a rich source of questions and possible answers to investigate.

The Transition Elements Suggestions

From these series of 'archetypal acts', a series of cross references have been identified, that provides the possible physical and spiritual activities of the transition elements.

The questions to ask are

A) Which Arm – the Internalised Etheric or World Physical?
B) Is it a Primary or Secondary planetary element?
C) Which 'energetic' ring is it on?
D) Which Venus pentagram and position?
E) Which 'Steiner' Arm is it associated with?
F) What is its 'major' element pair?

My suggestions follow. What are yours?

The Etheric Ring

ZINC Zn

Chemical Data: Atomic No - 30, Atomic weight 65.409, solid at 298K
Steiner Arm: Spirit +
Gyro Arm:
Internal Etheric / World Physical
Planetary element: Mercury 1
Streaming movement, the lymphatic system, sap flow

Major element relative: Gallium Etheric working into the World Physical
Venus Pentagram: A Physical force dominate Spirit - manifesting
Ag. Course Name:
Free Earth / Bound Chemical Ether

In nature: An essential element of nutrition, non toxic, High P blocks it, involved in protein formation where low, amino build up, leaf rossetteing occurs, bark brittle
Homeopathy: Restlessness, a need to keep moving and busy, repeating things, nervous exhaustion, prostrate problems.
Spiritual Activity: Facilitates the Etheric body working into the Physical processes to balance an over strong astral activity.

COPPER Cu

Chemical Data: Atomic No - 29, Atomic Weight - 63.546, solid at 298K
Steiner Arm: Etheric +
Gyro Arm: Internalised Etheric
Planetary element: Venus 1
Opens Etheric formative forces and nourishes what Mars thrust into space

Major element relative: Calcium (strong)
Activates the internalised Etheric
Venus Pentagram: B
Spirit dominating the physical elements - Being
Ag. Course Name: Bound Chemical Ether

In Nature: Conducts electricity and heat, often strongly bound to organic matter, more than other elements, Cu availability reduces as pH and Ca increases, influences CHO and nitrogen metabolism, needed for good pollen viability, aids reproduction in animals, Zn limits Cu, affinity for S, not found near Silicates, most common as $CuFeSO_4$, related to vitamin B, shellfish and mollusks use copper to breath rather than iron, lobster etc is high in Cu—holds proteineous tissue inside and expels Ca to the shell, they are water breathers, found in high amounts in the liver, pigmented areas and tumourous tissue, active metabolism leads to higher Cu, serum Cu high in schizophrenia, manic depression and epilepsy, high Cu gives resistance to cholera.
Homeopathy: (when short) Maintaining control, serious, hard working, sticking to the rules, Do not like criticism, tendency to cramps, In the metabolism it is necessary for the functioning of diverse proteins and oxidative enzymes. It is indispensable for the synthesis of haemoglobin, for the healthy functioning of nerves, and for the formation of bones. Moreover, it promotes the development of connective tissue for the heart's circulatory system.
Spiritual Activity: Opening up the organism (etheric body) to the workings of the astral and spirit.

NICKEL Ni

Chemical Data: Atomic No - 28, Atomic Weight - 58,693, solid at 298K
Steiner Arm: Astral +
Gyro Arm: World Etheric
Planetary element: Mars 1
The force which carries the spiritual archetype into the physical. The growth point

Major element relative: Potassium (strong)
Venus Pentagram: A
Physical force dominate
Spirit - manifesting
Ag. Course Name: Free Water

In Nature: too much Nickel makes you susceptibility to dermatitis and asthma should be increasing. In agriculture it is related to filling the grain, nixies are water spirits, magnetic, dissolves Carbon to make hard steel, Cu-esque but harder, Ni stored in liver & pancreas, high in Insulin and in grey hair, prepares for iron process, but can not replace Cu. Associated with Si, S and As but not O or CO. Used to hydrogenate Vegetable oils.

Homeopathy: Work hard and be nice, do not show much emotion, maintain harmony, fear of exams as they have done their best.

Spiritual Activity: Opens the World Etheric to the workings of the Astral and Spirit.

COBALT Co

Chemical Data: Atomic No - 27, Atomic Weight - 58.933, solid at 298K
Steiner Arm: Physical +
Gyro Arm: Internalised Spirit
Planetary element: Jupiter 1
Moulds 'plastic rounded forms' around Saturn's archetypal structures

Major element relative: Krypton (strong)
Venus Pentagram: B
Spirit dominating the physical elements - Being
Ag. Course Name: Bound warmth

In Nature: Vit B12 essential for ruminant N metabolism, low reproduction, found in igneous rock same distribution as Mg, shale's, Mn can block it, needed for symbiotic N2 fixation and rhizobial growth. Tougher than Ni, Ni > Mg, Co > Fe, Ni > plants, Co > animals, friends of As, S, Si, bi valent. Co salts used to oxidise linseed into varnish, found in liver pancreas and thymus gland, lack > pernicious anemia, polycythemia is an overproduction of red blood corpuscles = Ego does not control the bloods economy or the iron process in the blood, Co in thymus, which is a pre puberty gland, so Co paves the way for Fe in a 'Cu' phase. Used to intensify atom bombs and slow radioactive decomposition of uranium, involved in cyanogen processes.

Homeopathy: Adapted to neurasthenic spinal states. Sexual disturbances. Fatigue, agitation, and bone pains; worse in morning. Impotence, due to pressure, inadequacy and concern before beginning something.

Spiritual Activity: Facilitates the incarnation of the Spirit through the Astrality into the Etheric processes.

IRON Fe

Chemical Data: Atomic No - 26, Atomic Weight - 55.845, solid at 298K
Steiner Arm: Spirit –
Gyro Arm:
World Spirit / Internalised Spirit
Planetary element: Saturn 1
The spiritual archetype upon which life builds

Major element relative: Bromide (weak)
Venus Pentagram: A
Physical force dominate
Spirit - manifesting
Ag. Course Name:
Free Fire / Bound Warmth

In Nature: Carries oxygen and ego forces in the blood, helps cosmic weightless elements enter the sphere of gravity, enables us to anchor our personalities in our bodily processes, if low no 'presence of mind', close to Carbon, but has to balance S processes, Fe renders the cyanide processes harmless, has similarities to Pb in it ability to spontaneously combust when ground very fine. Carrier of the forces of embodiment > mummification, ability to absorb and retain formative forces, gives and takes oxygen easily, the breather metal, sensitive to light, negates poisoning processes of As and CN, bivalent form similar to Zn, Fe is light sensitive.

Homeopathy: Best adapted to young weakly persons, anæmic and chlorotic, with pseudo-plethora, who flush easily; cold extremities; over sensitiveness; worse after any active effort. Weakness from mere speaking or walking though looking strong. Pallor of skin, mucous membranes, face, alternating with flushes. Orgasms of blood to face, chest, head, lungs, etc. Irregular distribution of blood. Pseudo-plethora. Muscles flabby and relaxed.

Spiritual Activity: Carries the spirit into the controlling of the etheric activities behind physical processes.

MANGANESE Mn

Chemical Data: Atomic No - 25, Atomic Weight - 54.938, solid at 298K

Steiner Arm: Spirit –

Gyro Arm:
World Spirit / Internalised Astral

Planetary element: Saturn 2
Seed formation, fulfillment of karma in time

Major element relative:
Bromide (weak)

Venus Pentagram: B
Spirit dominating the physical elements - Being

Ag. Course Name:
Bound Light / Free Fire

In Nature: Activated thru oxidation processes, likes low pH, likes C (ala Fe), fiery nature, causes alcohol and ether vapors to combust, used in glass manufacture, Saturnian nature, sclerotic tendency, used as a drying agent in varnish and paints, salamanders linked to ripening processes. Human diets with too little manganese can lead to slowed blood clotting, skin problems, changes in hair colour, lowered cholesterol levels, and other alterations in metabolism. In animals, eating too little manganese can interfere with normal growth, bone formation, and reproduction, high levels of manganese dust in air may have mental and emotional disturbances, and their body movements may become slow and clumsy. manganese injures a part of the brain that helps control body movements, effects reproductive ability. Manganese is an antioxidant nutrient which is important in the blood breakdown of amino acids and the production of energy. Manganese is necessary for the metabolism of Vitamin B-1 & Vitamin E. This mineral activates various enzymes which are important for proper digestion and utilization of foods. Manganese is also a catalyst in the breakdown of fats & cholesterol, helps nourish the nerves & brain, is necessary for normal skeletal development, and maintains sex hormone production. A deficiency of manganese may result in paralysis, convulsions, dizziness, ataxia, loss of hearing, digestive problems, and blindness & deafness in infants.

Homeopathy: Inflammation of bones or joints, with nightly digging pains Asthmatic persons who cannot lie on a feather pillow. Syphilitic and chlorotic patients with general anæmic and paralytic symptoms often are benefited by this drug. Gout. Chronic arthritis. For speakers and singers. Great accumulation of mucus. Growing pains and weak ankles. General soreness and aching; every part of the body feels sore when touched; early tuberculosis.

Spiritual Activity: Directs the internal spirits working into physical processes.

CHROMIUM Cr

Chemical Data: Atomic No - 24, Atomic Weight - 51.996, solid at 298K	*Major element relative:* Selenium (strong)
Steiner Arm: Etheric	*Venus Pentagram:* A
Gyro Arm: Internalised Astral	Physical force dominate Spirit - manifesting
Planetary element: Jupiter 2	
Plant pharmacology, formation of oils, alkaloids & glycosides. Working of terrestrial light and warmth	*Ag. Course Name:* Bound Light

In Nature: Loves C, hardening a step towards sclerosis, Jupiter influences marked, similarities to tin, organises light and air, used as a mordant, sylphs, tanning and mummification very quick. Chromium is a mineral that works with insulin in the metabolism of sugar and stabilization of blood sugar levels. Chromium also cleans the arteries by reducing cholesterol & triglyceride levels; helps transport amino acids to where the body needs them; and helps control the appetite.

persons with low levels of chromium in their bodies aorta. Anxious pressure on whole chest. Fatty heart. Degenerative states, has brain softening. Atheroma of arteries of brain and liver.

Spiritual Activity: Engages astrality into etheric processes in body chemistry.

VANADIUM Va

Chemical Data: Atomic No - 23, Atomic Weight - 50.941, solid at 298K	*Major element relative:* Arsenic (strong)
Steiner Arm: Astral –	*Venus Pentagram:* B
Gyro Arm: World Astral	Spirit dominating the physical elements - Being
Planetary element: Mars 2	
Ordering of substance into starch and protein. Termination of living growth	*Ag. Course Name:* Free Light

In Nature: Makes very hard steels, can be found in the environment in algae, plants, invertebrates, fishes and many other species. In mussels and crabs vanadium strongly bio accumulates, which can lead to concentrations of about 105 to 106 times greater than the concentrations that are found in seawater. Vanadium causes the inhibition of certain enzymes with animals, which has several neurological effects. Next to the neurological effects vanadium can cause breathing disorders, paralyses and negative effects on the liver and kidneys.

Laboratory tests with test animals have shown, that vanadium can cause harm to the reproductive system of male animals, and that it accumulates in the female placenta. Vanadium can cause DNA alteration in some cases, but it cannot cause cancer with animals.

Homeopathy: Hesitating to put their talents into practice – weak Mars 2? oxygen carrier and a catalyser, hence its use in wasting diseases. Increases amount of haemoglobin, also combines its oxygen with toxins and destroys their virulence. Also increases and stimulates phagocytes. A remedy in degenerative conditions of the liver and arteries. Anorexia and symptoms of gastro intestinal irritation; albumen, casts and blood in urine. Tremors; vertigo; hysteria and melancholia; neuro-retinitis and blindness. Anæmia, emaciation. Cough dry, irritating and paroxysmal, sometimes with hæmorrhages. Irritation of nose, eyes and throat. Tuberculosis, chronic rheumatism, diabetes. Acts as a tonic to digestive function and in early tuberculosis. Arterio-sclerosis, sensation as if heart was compressed, as if blood had no room in the which task to choose.

Spiritual Activity: Etheric stimulation of physical processes.

TITANIUM Ti

Chemical Data: Atomic No - 22, Atomic Weight - 47.867, solid at 298K
Steiner Arm: Physical
Gyro Arm: Internal Physical
Planetary element: Venus 2
Excretion of what falls out of the life processes, eg cellulose in the rings of the tree or the potassium salts in the bark. Separates the substance from ether forces.

Major element relative: Germanium
Venus Pentagram: A
Physical force dominate
Spirit – manifesting
Ag. Course Name: Bound Life ether

In Nature: Low toxicity, not cancerous, used to treat ovarian cancer. Used as a sunscreen, not of nutritive value it appears.

Homeopathy: Fails to even start the job, Is found in the bones and muscles. Has been used in lupus and tuberculosis processes externally, also in skin disease, nasal catarrh, etc. Apples contain 0. 11 per cent of Titan. Imperfect vision, the peculiarity being that half an object only could be seen at once. Giddiness with vertical hemiopia. Also, sexual weakness, with too early ejaculation of semen in coitus. Bright's disease. Eczema, lupus, rhinitis.

Spiritual Activity: Stimulate the Etherics against the Astral.

SCANDIUM Sc

Chemical Data: Atomic No - 21, Atomic Weight - 44.955, solid at 298K
Steiner Arm: Spirit +
Gyro Arm:
World Physical / Internalised Physical
Planetary element: Mercury 2
Organ formation due the confluence of movement, wood out of the cambium

Major element relative:
Gallium (weak)
Venus Pentagram: B
Spirit dominating the physical elements – Being
Ag. Course Name:
Free Earth / Bound Life

In Nature: ScSo4 stimulates the germination of seeds, can be found in houses in equipment such as colour televisions, fluorescent lamps, energy-saving lamps and glasses. With water animals scandium causes damage to cell membranes, which has several negative influences on reproduction and on the functions of the nervous system. Considered a Rare Earth, found in Uranium minerals and Fe & Mg rocks and thortveitite. Easily oxides, used in lighting, light but high melting temperature.
Homeopathy: Looking and comparing, unsure of Opens Etheric formative forces and nourishes what Mars thrust into space.
Spiritual Activity: Etheric stimulation of physical processes.

The Astral Ring

CADMIUM Cd

Chemical Data: Atomic No - 48, Atomic Weight - 112.411, solid at 298K
Steiner Arm: Spirit +
Gyro Arm:
World Physical / Internal Etheric
Planetary element: Mercury 1
Streaming movement, the lymphatic system, sap flow

Major element relative: Indium
World Astrality into the W Physical
Venus Pentagram: A
Physical force dominate Spirit - manifesting
Ag. Course Name: Moon 1,
Mercury 1, / Mercury 2, Venus 2

In Nature: Found in combination with Zn, high ability to absorb neutrons, used in batteries & colouring agents, telephone cables, Cannabis Sativa has high levels of Cd, found in liver, mushrooms, shellfish, mussels, cocoa powder and dried seaweed. Cadmium is first transported to the liver through the blood. There, it is bonded to proteins to form complexes that are transported to the kidneys. Cadmium accumulates in kidneys, (and placenta) where it damages filtering mechanisms. This causes the excretion of essential proteins and sugars from the body and further kidney damage. It takes a very long time before cadmium that has accumulated in kidneys is excreted from a human body. In cigarettes and P fertiliser. Skeletal collapse due to interference in Ca metabolism, prostrate cancer, disturbed enzyme function.

Homeopathy: Repetition of the past, have been successful and on the way out, they know best, arrogant, stubborn as they know best, but nothing new to give. Its pathogenesis gives symptoms corresponding to very low forms of disease, as in cholera, yellow fever, where, with exhaustion, vomiting, and extreme prostration, the disease runs deathward. Important gastric symptoms. Carcinoma ventriculi; persistent vomiting.

Spiritual Activity: Drives the astrality into physical processes (see P) easily disrupting the etheric.

SILVER Ag

Chemical Data: Atomic No - 47, Atomic Weight - 107.868, solid at 298K
Steiner Arm: Etheric
Gyro Arm: Internalised Etheric
Planetary element: Venus 1
Major element relative: Strontium

Venus Pentagram: B
Spirit dominating the physical elements - Being
Ag. Course Name:
Mercury 2, Venus 2

In Nature: Universal healer, high conductivity, mirrors, ceaseless repetition and wave like reproduction, tends to be colloidal, all growth and body building processes, reproduction

Homeopathy: Emaciation, a gradual drying up, desire for fresh air, dyspnœa, sensation of expansion and left-sided pains are characteristic. The chief action is centered on the articulations and their component elements, bones, cartilages, and ligament's. Here the small blood vessels become closed up or withered and carious affections result. Holding onto a successful position, and the past, traditions. (Moon)

Spiritual Activity: Stimulating etheric forces touched by the astrality enough to direct them into manifestation.

PALLADIUM Pa

Chemical Data: Atomic No- 46, Atomic Weight - 106.42, solid at 298K
Steiner Arm: Astral
Gyro Arm: World Etheric
Planetary element: Mars 1
The force which carries the spiritual archetype into the physical. The growth point

Major element relative: Rb
Venus Pentagram: A
Physical force dominate
Spirit - manifesting
Ag. Course Name:
Mercury 1, Venus 1

In Nature: Like mercury, palladium is cytotoxic and kills or damages cells. Palladium also causes considerable damage and degradation of DNA and exacerbates hydroxyl radical damage Palladium also damages cell mitochondria and inhibits enzyme activity and function, highly mobile and toxic, used in dentistry, Disturbance of collage synthesis like bone and cartilage; Obstruction of thymidin in the

DNA; Accumulation in body organs; blocks the action of a number of enzymes and interferes with use of energy by nerves and muscles; induces lung malfunction and produces abnormal foetuses.

Homeopathy: An ovarian remedy; produces the symptom-complex of chronic Oophoritis. Useful where the parenchyma of the gland is not totally destroyed. Acts also on mind and skin. Motor weakness, averse to exercise. Mind.-Weeping mood. Love of approval. Pride; easily offended. Inclined to use violent language. Keeps up brightly when in company, much exhausted afterwards, and pains aggravated.

Spiritual Activity: Astral stimulation of the organising Etheric processes.

RHODIUM Rh

Chemical Data: Atomic No - 45, Atomic Weight - 102.90, solid at 298K
Steiner Arm: Physical
Gyro Arm: Internal Spirit
Planetary element: Jupiter 1
Moulds 'plastic rounded forms' around Saturn's archetypal structures.

Major element relative: Xenon
Venus Pentagram: B
Spirit dominating the physical elements - Being
Ag. Course Name:
Jupiter 2, Saturn 2

In Nature: Rhodium has a higher melting point and lower density than platinum. It has a high reflectance and is hard and durable. Upon heating it turns to the oxide when red and at higher temperatures turns back to the element. highly toxic and a carcinogenic, automobile exhaust, detoxification, Supplemental rhodium increased the hematocrits and liver oxidative ability of both nickel-deficient and -supplemented chicks, and increased total liver lipids, liver lipid phosphorus, and liver cholesterol in the nickel deficient chicks alone. Rhodium did not increase the signs of nickel deficiency.

In Homeopathy: Nervous and tearful. Frontal headache; shocks through head. Fleeting neuralgic pains in head, over eyes, in ear, both sides of nose, teeth. Loose cold in head. Lips dry. Nausea especially from sweets. Dull headache. Stiff neck and rheumatic pain down left shoulder and arm. Itching in arms, palms and face. Loose stools with gripings in abdomen. Hyper-active peristalsis, tenesmus after stool. More urine passed. Cough scratchy, wheezy. Thick, yellow mucus from chest. Feels weak, dizzy and a tired feeling.

Spiritual Activity: Astral stimulation of the Ego and nerve sense pole.

RUTHENIUM Ru

Chemical Data: Atomic No - 44, Atomic Weight - 101.07, solid at 298K
Steiner Arm: Spirit –
Gyro Arm:
World Spirit / Internal Spirit
Planetary element: Saturn 1
The spiritual archetype upon which life builds

Major element relative: Iodine
Venus Pentagram: A
Physical force dominate Spirit - manifesting
Ag. Course Name: Jupiter 1 Saturn 1, Jupiter 2, Saturn 2

In Nature: Ruthenium is found as the free metal, sometimes associated with platinum, osmium and iridium, in North and South America, and in South Africa. There are few ores is also associated with nickel and deposits, highly toxic and as carcinogenic, is retained strongly in bones, hard brittle durable and corrosion resistant, strong affinity for hydrogen, Ruthenium 103 used in nuclear bombs.

Homeopathy: A heavy task, lots of work they must get done, resolute, inspired.

Spiritual Activity: The inspiration for the long process of creation, astral inspired spirit.

TECHNETIUM Tc

Chemical Data: Atomic No - 43, Atomic Weight - 98, solid at 298K
Steiner Arm: Spirit –
Gyro Arm:
World Spirit / Internal Astral
Planetary element: Saturn 2
Seed formation, fulfillment of karma in time

Major element relative: Iodine, Manganese
Venus Pentagram: B
Spirit dominating the physical elements - Being
Ag. Course Name: Jupiter 1, Saturn1 / Jupiter 2 Mars 2

In Nature: Radioactive and not naturally occurring, which is unusual for such a light element.

Homeopathy: Practising creativity > channeling creativity.

Spiritual Activity: World Astral into the World Spirit.

MOLYBDENUM Mo

Chemical Data: Atomic No - 42, Atomic Weight - 95.94, solid at 298K

Steiner Arm: Etheric

Gyro Arm: Internal Astral

Planetary element: Jupiter 2
Plant pharmacology, formation of oils, alkaloids & glycosides. Working of terrestrial light and warmth

Major element relative: Tellurium

Venus Pentagram: A
Physical force dominate
Spirit – manifesting

Ag. Course Name: Jupiter 2, Mars 2

In Nature: Discovered in a Pb ore, metal is greasy and used as a lubricant, high melting point, resistant to acids, electrodes for electrically heated glass furnaces and fore heaths. The metal is also used in nuclear energy applications and for missile and aircraft parts. Molybdenum is valuable as a catalyst in the refining of petroleum. It has found applications as a filament material in electronic and electrical applications. Molybdenum is an essential trace element in plant nutrition, yet highly toxic. evidence of liver dysfunction with hyperbilirubinemia has been reported in workmen chronically exposed in a Soviet Mo-Cu plant, signs of gout have been found in factory workers and among inhabitants of Mo-rich areas of Armenia, joint pains in the knees, hands, feet, articular deformities, erythema, and edema of the joint areas. SO_4 limits Mo uptake, PO_4 enhances uptake, Mo important in the enzymes nitrogenase and nitrogen reductase, works with Fe @ 9:1, when low can lead to N deficiency, problem in Si soils, low pH and even peat. Cucurbits & legumes high Mo need, liming helps. High Mo > Cu deficiency.

Homeopathy: The beginning of expressing ones own creativity. Got the inspiration and need a little more will.

Spiritual Activity: Astral stimulant.

NIOBIUM Nb

Chemical Data: Atomic No - 41, Atomic Weight - 92.906, solid at 298K

Steiner Arm: Astral

Gyro Arm: World Astral

Planetary element: Mars 2
Ordering of substance into starch and protein. Termination of living growth

Major element relative:
Antimony (strong)

Venus Pentagram: B
Spirit dominating the physical elements - Being

Ag. Course Name: Mars 1, Jupiter 1

In Nature: Occurs in a mineral columbite. Formerly known as colombium (Cb). used in stainless steel alloys for nuclear reactors, jets, missiles, cutting tools, pipelines, super magnets and welding rods. Niobium, when inhaled, is retained mainly in the lungs, and secondarily in bones. It interferes with calcium as an activator of enzyme systems. In laboratory animals, inhalation of niobium nitride and/or pentoxide leads to scarring of the lungs at exposure levels of 40 mg/m³.

Homeopathy: Hesitation and doubt about showing creativity. Difficult to decide which chosen path to follow, lots of unfinished projects.
Spiritual Activity: Astral stimulant.

ZIRCONIUM   Zr

Chemical Data: Atomic No - 40, Atomic Weight - 91.224, solid at 298K
Steiner Arm: Physical
Gyro Arm: Internal Physical
Planetary element: Venus 2
Excretion of what falls out of the life processes, eg cellulose in the rings of the tree or the potassium salts in the bark. Separates the substance from ether forces.

Major element relative: Tin
Venus Pentagram: A
Physical force dominate
Spirit - manifesting
Ag. Course Name: Moon 2 Mercury 2

In Nature: Used in alloys such as zircaloy, which is used in nuclear applications since it does not readily absorb neutrons, used in catalytic converters, percussion caps and furnace bricks, lab crucibles. have low systemic toxicity. Zirconium 95 is one of the radionuclides involved in atmospheric testing of nuclear weapons. It is among the long-lived radionuclides producing increased cancers risk for decades and centuries to come.
Homeopathy: The first job, opening of ones own practice. Lots of ideas and want to show their value.
Spiritual Activity: Putting astral inspiration into practical applications. Astral stimulation of the physical body.

YTTRIUM Y

Chemical Data: Atomic No - 39, Atomic Weight - 88.905, solid at 298K
Steiner Arm: Spirit +
Gyro Arm:
World Physical / Internal Physical
Planetary element: Mercury 2
Organ formation due the confluence of movement, wood out of the cambium

Major element relative: Indium
Venus Pentagram: B
Spirit dominating the
physical elements - Being
Ag. Course Name: Moon 1
Mercury 1 / Moon 2 Mercury 2

In Nature: Colour televisions, fluorescent lamps, energy-saving lamps and glasses. rarely be found in nature, as it occurs in very small amounts. Yttrium is usually found only in two different kinds of ores. cause lung embolisms, cancer and accumulates in the liver. With water animals yttrium causes damage to cell membranes, which has several negative influences on reproduction and on the functions of the nervous system.

Homeopathy: Exploring creative abilities, unsure and don't trust their own inspirations

Spiritual Activity: Bringing astrality to the physical realm.

The Spirit One Ring

MERCURY Hg

Chemical Data: Atomic No - 80, Atomic Weight - 200.59, liquid at 298K

Steiner Arm: Spirit +

Gyro Arm: Internal Etheric / World Physical

Planetary element: Mercury 1 Streaming movement, the lymphatic system, sap flow.

Major element relative: Ti

Venus Pentagram: A Physical force dominate Spirit - manifesting

Ag. Course Name: Leo

In Nature: Mostly found in Europe as HgSO4, found with Ag, Cu, Sb, strongly linked to S, maintains its own 'liquid' form, where there are congestions Hg dissolves these and brings movement into stagnation. Highly toxic, moves freely through the body and is easily past from mother to child, HgCN is used for diphtheria,

Homeopathy: Takes hold of life processes and guiding it to a higher level, separating process that must be reformed into the organism. Slow in answering questions. Memory weakened, and loss of will-power. Weary of life. Mistrustful. Thinks he is losing his reason.

Spiritual Activity: Internal Spirit directs the forming of substance.

GOLD Au

Chemical Data: Atomic No - 79, Atomic Weight - 196.966, solid at 298K

Steiner Arm: Etheric +

Gyro Arm: Internalised Etheric

Major element relative: Barium

Planetary element: Venus 1 Opens Etheric formative forces and nourishes what Mars thrust into space

Venus Pentagram: B Spirit dominating the physical elements - Being

Ag. Course Name: Gemini

In Nature:

Homeopathy:

Spiritual Activity: Internal Spirit directs the Internal Etheric Body.

PLATINUM Pt

Chemical Data: Atomic No - 78, Atomic Weight - 195.078, solid at 298K
Steiner Arm: Astral +
Gyro Arm: World Etheric
Major element relative: Cesium
Planetary element: Mars 1
The force which carries the spiritual archetype into the physical. The growth point.

Venus Pentagram: A
Physical force dominate
Spirit - manifesting
Ag. Course Name: Virgo

Spiritual Activity: Cosmic Spirit directs the World Etheric.

IRIDIUM Ir

Chemical Data: Atomic No - 77, Atomic Weight - 192.21, solid at 298K
Steiner Arm: Physical +
Gyro Arm: Internalised Spirit
Major element relative: Radon

Planetary element: Jupiter 1
Moulds 'plastic rounded forms' around Saturn's archetypal structures.
Venus Pentagram: B
Spirit dominating the physical elements - Being
Ag. Course Name: Pisces

In Nature:
Homeopathy:
Spiritual Activity: Internal Spirit

OSMIUM Os

Chemical Data: Atomic No - 76, Atomic Weight - 190.23, solid at 3306K
Steiner Arm: Spirit –
Gyro Arm:
World Spirit / Internal Spirit
Venus Pentagram: A
Physical force dominate
Spirit - manifesting

Major element relative: At
Planetary element: Saturn 1
The spiritual archetype upon which life builds
Ag. Course Name: Sagittarius

In Nature:
Homeopathy:
Spiritual Activity: Cosmic Spirit directs the World Spirit into incarnation.

RHENIUM Re

Chemical Data: Atomic No - 75,
Atomic Weight - 186.20, solid at 3459K
Steiner Arm: Spirit –
Gyro Arm:
World Spirit / Internal Astral
Venus Pentagram: A
Physical force dominate
Spirit - manifesting

Major element relative: At
Planetary element: Saturn 2
Seed formation, fulfillment of
karma in time
Ag. Course Name: Pisces

In Nature:
Homeopathy:
Spiritual Activity: Cosmic Spirit directs the World Spirit to bring conclusion.

TUNGSTEN W

Chemical Data: Atomic No - 74,
Atomic Weight - 183.84, solid at 3695K
Steiner Arm: Etheric
Gyro Arm: Internal Astral
Venus Pentagram: B
Spirit dominating the
physical elements - Being

Major element relative: Polonium
Planetary element: Jupiter 2
Plant pharmacology, formation of oils,
alkaloids & glycosides. Working of
terrestrial light and warmth
Ag. Course Name: Scorpio

In Nature:
Homeopathy:
Spiritual Activity: Cosmic Spirit directs the Internal Astral to bring success.

TANTALUM Ta

Chemical Data: Atomic No - 73,
Atomic Weight - 180.94, solid at 3290K
Steiner Arm: Astral –
Gyro Arm: World Astral
Venus Pentagram: A
Physical force dominate
Spirit - manifesting

Major element relative: Bismuth
Planetary element: Mars 2
Ordering of substance into starch and
protein. Termination of living growth
Ag. Course Name: Aries

In Nature:
Homeopathy:
Spiritual Activity: Cosmic Spirit directs the World Astral.

HAFNIUM Hf

Chemical Data: Atomic No - 72, Atomic Weight - 178.49, solid at 2506K
Steiner Arm: Physical
Gyro Arm: Internal Physical
Venus Pentagram: B
Spirit dominating the physical elements - Being
Major element relative: Lead

Planetary element: Venus 2
Excretion of what falls out of the life processes, eg cellulose in the rings of the tree or the potassium salts in the bark. Separates the substance from ether forces.
Ag. Course Name: Cancer

In Nature:
Homeopathy:
Spiritual Activity: Cosmic Spirit directs the Internal Physical.

LANTHANUM La

Chemical Data: Atomic No - 57, Atomic Weight - 174.96, solid at 1936K
Steiner Arm: Spirit +
Gyro Arm:
World Physical / Internal Physical
Major element relative: Ti

Planetary element: Mercury 2
Organ formation due the confluence of movement, variation in individual form for the circumstance
Venus Pentagram: A
Physical force dominate Spirit - manifesting

In Nature:
Homeopathy:
Spiritual Activity: Cosmic Spirit directs the World Physical to birth.

And so on...

References

for this section have come from

- *R. Hauschka – Nature of Substance*
- *W. Pelikan – Secrets of Metals*
- *J. Scholten – Homoeopathy and the Elements*
- *Hausemann and Wolff – The Anthroposophical Approach to Medicine Vol 2*
- *R Steiner – Spiritual Science and Medicine*
- *W Boericke – Homeopathic Materia Medica*
- *Lenntech – website through Google*

The Lanthanides and Actinides

The 7th Harmonic – The Spirit's Toolbox

These two sets of elements exist in the outer two circles of the periodic table. The Lanthanides are elements 57 – 70 and are placed between Barium and Lutetium. (This can vary depending on which periodic table used as reference).

The Actinides are in the outer circle and are elements 89 – 102 and placed between Radium and Lawrencium.

The Lanthanides are called the Rare Earths, with only two of their number being radio active, while the Actinides are nearly all radioactive, with the majority only existing as debris of nuclear reactors or bombs.

Within each group, the elements are considered to be very similar, in quality, however there are some differences to be seen in their melting points and significantly in their uses. They have an interesting character of their radii size diminishing as the atomic weight increases, while their valence order presents an picture of two halves.

The Sixth Ring

On the gyroscopic diagram, there are six primary layers and each can be divided into two parts however, it is the sixth layer, the 12 fold zodiac related layer, that has its two layers more definitely defined. When looking at each of the layers as part of a vortex form, the 12 fold layer is at the top, and forms the most outer part of the vortex. For the same amount of vertical distance in the vortex, (or relative .618 ratios) as the other layers, there is a stretching and

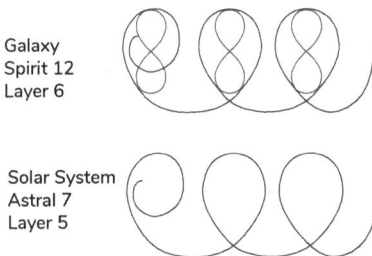

Galaxy
Spirit 12
Layer 6

Solar System
Astral 7
Layer 5

bending activity taking place, as the vortex reaches the top and begins to curve around to make a sphere. This allows for the two inner layers to become manifest. This ordering is mirrored in the planetary rulerships of the zodiac. Each planet – of the primary six planets – rules two signs.

This sixth ring is the ring of spiritual activity, and contains most of the radioactive elements and the Rare Earths. One of their qualities is they have a 'loose' outer ring of electrons, that freely gives and takes electrons, as the environment require. They act as the 'chaotic adjusters' as needed, allowing spirit to work miracles.

To Lemniscate or Not

There is the question of whether these elements should be put through a lemniscate loop, before enfolding them back over the gyroscope. This was done for the Transition Elements (p. 97), because they are elements intimately woven into life processes, which is often accompanied by a lemniscate process. The sequence of hardness and melting points – Ca, Zn > Sc, Ga – fits if a lemniscate twist is made, and so on. With the Lats and Acts, they are NOT elements intimately associated with life processes. If anything they are dangerous to life processes. They also do not show any reversing of melting points and hardness as seen in the Transition elements. So I have concluded my initial decision to lemniscate these elements was wrong, and I have adjusted my diagrams accordingly.

Where to begin the Group

There is some debate as to where the beginning and end of the series is. **Which elements sit in the 3B column of the transition elements** under Sc and Y? Is it La and Ac or Lu and Lr? From my various readings I have gone with La and Ac being the 'Transition Elements'. This leaves 14 groups of elements starting with Ce and Th and ending with Lu and Lr.

The Planets

With 14 groups of elements we can find a relationship to the planets. The planetary pattern we have used throughout these endeavors has three outer planets polarised against three inner planets with the Sun holding the central position. 14 = 2 x 7, which suggests we can apply this 7 fold pattern to these elements, as a possible reference. In this process, we can identify one side of the circle as being phase 1 'Being' elements, while the last half of the elements would be phase 2 'Manifestation' elements.

To find more indications for the activity of these elements we can 'play the same games' we have done with the other elements. Associations arise when these elements are placed over the energetic activities.

With the planetary associations, it is also possible to identify the elements relationship to the zodiac, and from this we have a hint of how each elements relates to the physical systems, Nerve Sense, Rhythmic or Metabolic. This grouping may save the need for potentising the various elements for the specific physical system. Instead use the one element from the group of three for the physical system you wish to address, with the energetic activity that needs support.

The practical uses of these elements further defines their qualities. The Etheric elements are used as catalysts often in glass making, the Spirit elements are used as neutron accumulators in nuclear reactions, the Astral elements are used in lasers, while the Physical elements are used in optics or not at all.

Homeopathy (Scholten) have done provings for these elements, which indicate they are used for 'spiritual illnesses', feelings of life or death, power issues, both too much and too little, and feelings of disintegration. With this being the Cosmic Spirit ring it also suggests 'spiritual maturity', intuitive capacities and feelings of oneness with creation. My own provings of the Lanthanide elements suggest they all have a very centering quality, with varying degrees of clarity and willingness to take action, according to their energetic group.

The problems of these elements radioactivity can be overcome by collecting them from a circle, using the method outlined on page 182. You then have the energetic resonance without the problems of their physical mass. These can be further potentised as you wish.

The available references for this group becomes

    A) **The Spiritual Ring**
    B) **Which 7 fold planet does it resonate with?**
    C) **Which gyroscopic energetic body arm is it in?**
    D) **What are the other major and transition elements it is near?**
    E) **What physical system does it relate to?**
    F) **The Journeys**

The Elements Effects

Given these elements manifest on the sixth ring of the periodic table, they are elements of the Cosmic Spirit sphere, of the second vertical plane of the Gyroscope. The primary plane is the Major elements and vibrates on the fourth harmonic. (see The 3D Periodic Table) So while this is a Spirit band, it is not the same resonance, given it is a seventh harmonic resonance, as the Spirit activity found in the Majors plane. Scholten contends these elements are related to working of the internal Spirit, and I agree. Using the 'polarity principle' of inner and outer, if the Majors are the basis of how the elements work in the outer world, these 'other' elements can be related to the Internal Spirits activity, upon life. Experience confirms this.

This suggests these elements help with how the internal Spirit can work with the other energetic activities. This is important for our present age as on the

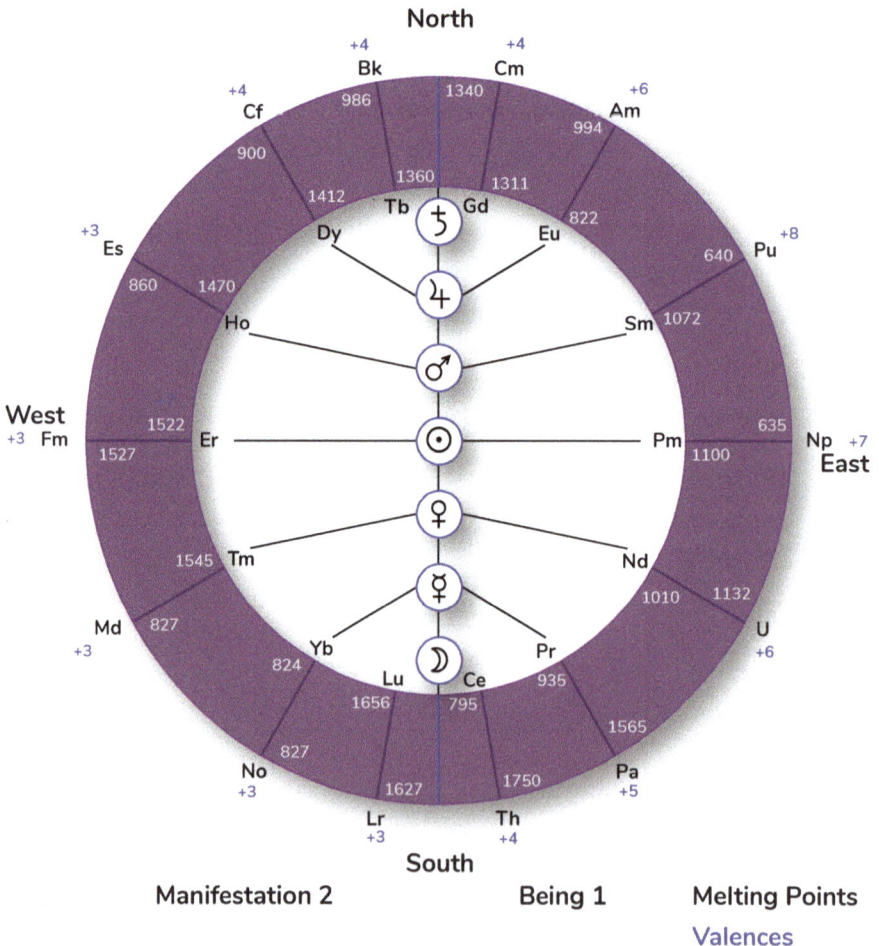

North

+4 Bk 986	+4 Cm 1340	
+4 Cf 900		+6 Am 994
	1360 Tb Gd 1311	
1412	♄	822 Eu
+3 Es 860	Dy	+8 Pu 640
1470 Ho	♃	Sm 1072
West +3 Fm 1522 1527	♂ Er	635 Np +7 East
	☉	Pm 1100
	♀	Nd
1545 Tm	☿	1010 1132
Md +3 827	Yb 824 Lu Ce Pr	U +6
	1656 795 935	1565
827 No +3	1627 1750	Pa +5
	Lr +3 Th +4	

South

Manifestation 2   Being 1   Melting Points

Valences

one hand we are evolving through a period where the spirit is bringing us more consciousness, however we are also being assaulted by many toxic substances, such as Chlorine and Fluoride that are bringing the World Spirit into us more strongly, and thus pushing out our internal Spirit processes. Many illnesses are based upon this circumstance. This is evident in the multitude of inflammatory and auto immune diseases, whose cause is the Etheric body loosing contact with the guiding plans of the Spirit. The Spirit often needs to be recombined with the other bodies, for most modern humans. Not only is this essential for health, due it having the architectural plans to your existence, but it is also essential if any real conscious freedom is to be achieved. Conscious Freedom occurs when the Spirit is in its right place, having the other bodies doing its bidding. Usually the Spirit is flung off center and becomes 'captured' by the Astrality. It runs its 'madness program' and the Spirit obsesses over it. The Spirit needs to be strengthened, so it stands

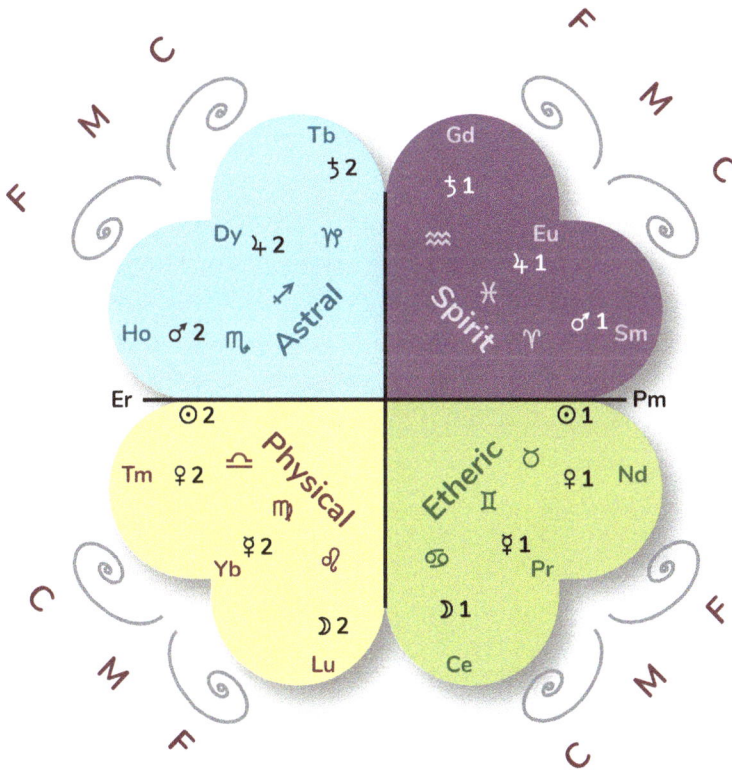

Planets
Constellations
Activities

C Metabolic
M Rhythmic
F Nerve Sense

in its own place, at the centre of the Astral whirlwind. The calm place is in the center and the strengthening of the Spirit will take you there. For health though it does need to 'wind into' all the other bodies. The archetypal journey, will do this for you.

The Journeys

This journey has two parts and uses the Lanthinides elements as 'experience strengtheners'. You can have the same experience, if you walk the journey around any Earth Circle.

There are two ways these elements can be approached. A significant body of work on the Lanthanides has been done by the Scholten school of homeopathy. His system is based on a sequential unfolding of 'a process' through the elements, ordered as a Flat Plane Spiral. The process has an expansion phase, a culmination followed by a waning phase. This can be consider 'The **Archetypal Journey**'. It starts at the beginning of the series and runs round the atomic weight order.

 His interpretations of the elements are framed within a psychological and physical reference. Based firstly upon 'the system' and then 'proven' by clinical results, observed with this reference in mind. The literature on this is readily available on the internet. This is very valid and valuable work, as far as it goes. This 'movement' is part of the classical homeopathic movement, and so is framed within that melieu, and expresses the limitations inherent in that approach. Scholten tells this story from its most dramatic angle. His emphasis is on the extremes of expansion and the inevitable failure in complete desperation, which does manifest in the very sick person. Looking at his story within the Glenopathic context, softens the sharp edges of this very fundamental process of normal growth.

I have done two provings of these elements, one starting at Cerium going anti-clockwise round to Lu, (as outlined above) which is the Archetypal path. The personal 'proving' of this journey is probably best to do, before 'the Manifestation journey'. The aim of the Archetypal journey is to enhance the natural process of the Internal Spirit integrating into the other bodies. This is a softer and more natural process, to start with than what is experienced doing the Manifest journey. By beginning at the Moon 1 position, the 'child' phase, and its first three elements are through the 'soft' Etheric. The Spirit merges more strongly with the Internal Etheric, and this cushions your journey through the Internal Spirit zone, where it reaches its culmination, before being challenged and moulded by the world in its journey through the Astrality. All this maturity is then solidified as it moves through the Internal Physical.

The Spirit is a strong and hostile 'physical' force. In the '**Manifest journey**', we start with the Spirit influence, moving through the Spirit zone. This hardening influence at the start of a persons 'Glenopathic journeying', could be challenging and counter productive, by hardening existing 'etheric errors'. The Etheric is were the 'patternings' of the growth forces reside, and they often need strengthening. When these are 'sick' the physical organism can not be sustained. So align the bodies, starting with the Etheric first, while using a natural cycle, which mirrors the manifesting processes found within a Moon cycle, will provide an easier start, to an amazing experience.

The **Manifest Journey** starts at Gd and goes clockwise around to Tb. This journey can be done through a whole moon cycle, beginning at new moon. The aim of this process is to move through the 'creation cycle' of the chemical elements, within a moon cycle. Astrology knows the power of the moon cycle, as a physical reality in the process of manifestation, at all natural levels. Dr Steiner (and elaborated by Dr Lievegoed) outlined the big picture of how manifestation occurs as a result of a Star force, moving through our Solar system spheres, followed by a journey through the Atmosphere, before we receive it here on Earth. Something is born at Moon 2, lives its life, as an expression of the Sun and 6 planets, and dies, at the Saturn 2 position. This truly is a Spirit journey. Each element is taken up to twice daily for two days each. Beginning at the New Moon. You can have a project in mind, or just go on the journey as it unfolds.

The Gyroscope

Another approach is to see the Lant. elements within the context of the gyroscopic organisation, and the energetic activities outlined within it. This goes some way to providing the behind the scenes activity, for what we find described by Scholten. When we see how the two perspectives relate to each other, we have the energetic body pattern of their psychological expressions.

The Etheric Group – Gathering the Life Forces

CERIUM Ce

Chemical Data: Atomic No – 58
Atomic weight 140.1
Ring: 6 – Cosmic Spirit
Planet: Moon 1 / Cancer
(a) Through reproduction on a small scale (cell division) and on a large scale (reproduction), creates a small and large chaos, into which the Saturn forces can impress their seal, and where in every cell the Spiritual archetype can be received anew.
Energetic Arm: World Physical
Physical Mode:
Cardinal, Metabolic system
Similar elements: Ti, Hg

The Archetypal Journey: Stage 1
(a) Feeling of not being in contact with the world and doing things without trial and error, but practicing with their mind.
(b) Temporary plans develop, but usually change before any significant efforts have begun
The Manifest Journey: Stage 7
The 'baby phase' is apparent, which shows the archetypal journey phase, however in this process this is the last minute touches needed before the show begins. Birth / product launch is just around the corner. Everything that can be done has been done and there is a faith that all will be well. The 'soft delicate vibe' found here encompasses everything that becomes 'stage fright'., waiting for the show to begin.

PRASEODYMIUM Pr

Chemical Data: Atomic No – 59,
Atomic weight 140.9,
Ring: 6 – Cosmic Spirit
Planet: Mercury 1 / Gemini
(a) Brings the semi fluid world of life into streaming movement, adapting itself to the chance conditions on Earth, changing the rigid Jupiter forms into forms adapted to what is possible.
Energetic Arm: Internal Etheric
Physical Mode: Mutable,
Rhythmic system
Similar elements: Ba

The Archetypal Journey: Stage 2
(a) Hesitant, not daring to do what they feel their own path. Keep looking from which side problems can arise.
The Manifest Journey: Stage 6
The Internal Etheric rises, and with it comes a reduction in 'thinking strength' but a increase in 'intuitive strength'. A soft feeling into things, that is putting resources into action, as a prototype of the real thing is being tested. 'Naive mistakes' are being looked for. Launch date / birth is not far away.
The Etheric can bring a sense of well being and health, or a degree of tiredness, initially. The driving Spirit and Astrality are pushed off. This is good. Sleep if you need

to, you are recuperating.

Inflammatory illnesses arise when the Spirit no longer properly directs the Etheric. This remedy is a cornerstone remedy for many of these 'illnesses of our time'.

NEODYMIUM Nd

Chemical Data: Atomic No – 60, Atomic weight 144.2
Ring: 6 – Cosmic Spirit
Planet: Venus 1 / Taurus
(a) Opens the Etheric Formative Forces into a cup or chalice and nourishes what Mars thrusts into space.

Energetic Arm:
Internal Etheric / World Etheric
Physical Mode:
Fixed / Nerve Sense system
Similar elements: Au

The Archetypal Journey: Stage 3
(a) They have to prove they can do it, by conquering their fear.
(b) Anxiety with new challenges; a sense of sink or swim, that one is forced by circumstances to act even though ones preparation feels incomplete.

The Manifest Journey: Stage 5
The conviviality element – This has a softness it gets from the Etheric activity, but a movement it gains from being Cardinal. After the hardness of the Spirit elements and the indecision of Pm, this is a very pleasant cushioning, that allows one to relax into artistic and social situations.

The goal is clarified, the odds have been weighed, and now you look for the resources, and the support of friends and associates to move your project along. There is a gentle opening of soul that allows you to 'go out' and be open to the support you need, and to meet it with friendliness.

Sun 1 – All things Considered

PROMETHIUM Pm

Chemical Data: Atomic No – 62, Atomic weight 145,
Ring: 6 – Cosmic Spirit
Planet: Sun 1

Energetic Arm: World Etheric
Physical Mode: All
Similar elements: Pt, Cs

The Archetypal Journey: Stage 4
Learning, training, wanting assurance they are doing OK.

The Manifest Journey: Stage 4
Scholten felt this remedy had no useful applications. It has an unusual characteristic of physically manifesting only briefly as a result of a nuclear reaction.

As the Sun 1 element we can sense a synthesising quality that leads to an assessment of what has occurred so far. Being part of the coming into being process, there is an unconsciousness that requires some faith still. There is a sense of purpose, but with a naivety and a slight egotism. I am not able, or allowed to fully comprehend what is really at stake.

I was invigorated, however slightly agitated, due to dealing with things from several different directions. There is an attempt to find the common theme and integrate all the parts, but nothing fully condenses. Feeling like I am sitting in the middle of something, and being softly buffered from all sides. I need to continue on with some faith.

The Spirit Group – Defining the Focus

SAMARIUM Sm

Chemical Data: Atomic No – 62, Atomic weight 150.3
Ring: 6 – Cosmic Spirit
Planet: Mars 1
(a) What has thus been created is placed forcefully into the world of space and now becomes visible in growth.

Energetic Arm:
World Etheric, Internal Spirit
Physical Mode:
Cardinal, Metabolic system
Similar elements: Ir

The Archetypal Journey: Stage 5
(a) Feeling the pressure of the task, perseverance and endurance needed for the large task ahead.
Becoming aware of the task ahead and working with persistence towards its manifestation.
The Manifest Journey: Stage 3
A calm centredness that is functional without the 'hardness' of Gd. Getting on with the jobs that need doing. Prepared to put in whatever hours are needed. The Spirit in the metabolic function also shows as a slight constipation.
Periodic 'hot flushes'.
Uses: If you are having trouble getting going on a project. Worked well for menopausal hot flushes, where the metabolism benefits from some direction from the Spirit, with Mars' impulsive bursts.

EUROPIUM Eu

Chemical Data: Atomic No – 63, Atomic weight 151.9,
Ring: 6 – Cosmic Spirit
Planet: Jupiter 1
(a) Jupiter forces round off, and in plastic beauty play around these severe Spirit forms, creating according to sublime and grand patterns.

(b) This phase takes the initial impulse from Saturn 1 and adapts it to the needs of the time and the environment in which it is to manifest
Energetic Arm: Internal Spirit
Physical Mode:
Mutable, Rhythmic system
Similar elements: Rn

The Archetypal Journey: Stage 6

(a) The confidence of things being achieved, and almost there.

(b) No longer needing to test one's strength; what one lacks in capacity is compensated for in terms of sheer perseverance. A sense of 'crawling through the desert' to reach ones goal.

The Manifest Journey: Stage 2

The Mutable elements of all the groups, appear to have a calm centered clarity, that suggests a synthesis of the elements on either side of them, occurs here. I suspect these will be the most useful remedies of each group.

This element generally brings in the Spirit to confidently focus on what needs doing. The internal Spirit is associated with the noble gases, and their self contained quality makes this the arm of the autism spectrum. The self contained quality of these three elements should be seen within this context. They are not withdrawn, but I noticed a reduced tolerance of superficiality and worldliness, during these days.

Uses: This is a very useful remedy, and works well to dampen down overly active astrality, that manifests as emotional and physical agitation, shaky movements, stilted speech, and general psychological disturbances. The Spirit gently pushes the Astral back into place.

GADOLINIUM Gd

Chemical Data: Atomic No – 64, Atomic weight 157.3,

Ring: 6 – Cosmic Spirit

Planet: Saturn 1 – Aquarius

(a) From the cosmic distances the Spirit works inwards and contracts, leading to the impress as of a seal into the Physical, a process reaching as far as crystallisation.

(b) This is where the 'star' impulse and intention for incarnation enters the astral sphere. The goal is clarified but the details of action are still to be finalised.

Energetic Arm: World Spirit

Physical Mode:
Fixed / Nerve Sense system

Similar elements: Os

The Archetypal Journey: Stage 7

At the summit, success, everything is OK

The Manifest Journey: Stage 1

The strong Spirit dominance of this element brings a clarity and centeredness, while the Fixed influence makes for a strong mental and rational tendency, that allows for a sense of objectivity. However the Saturn 1 influence is still unclear as to the details of the task ahead. This is a clarifying stage where the goals are defined, and strategic plans are begun.

Uses: Where the thinking needs to be clarified. Could be useful against headaches.

The Astral Group – Adjusting to the World Reactions

TERBIUM Tb

Chemical Data: Atomic No – 65, Atomic weight 158.9, *Ring:* 6 – Cosmic Spirit

Planet: Saturn 2 – Capricorn
Energetic Arm: World Spirit
Similar elements: Re

Physical Mode: Cardinal / Metabolic system
The Archetypal Journey: Stage 8
(a) Maintaining the success, but aware of the possibility of loosing it.
(b) Full confidence in ones inner capacities and talents (but nevertheless thwarted in their full outer expression).
The Manifest Journey: Stage 14
The complete person. The job is done, success has been achieved, and there is nothing more to be do. A bit wandering around with nothing to do. Feeling mature, centered, standing tall.
And now the responsibility for what to do with the achievement. What next?

DYSPROSIUM Dy

Chemical Data: Atomic No – 66, Atomic weight 162.5, *Ring:* 6 – Cosmic Spirit
Planet: Jupiter 2 (a) has been described as chemist, serving mans movement in the muscles. This chemistry while serving movement is a destructive process, that leads to the aroma of flowers. Sugar becomes glucosides, carbohydrates become ethereal oils and proteins alkaloids.
(b) Late 40s sense of optimism about ones skills, achievements and place in the world.
Energetic Arm: Internal Astral
Physical Mode:
Mutable, Rhythmic system
Similar elements: Po

The Archetypal Journey: Stage 9
(a) Need to fight to maintain position, overdoing things, boasting and inflating, fighting for the cause
(b) Anxiety with respect to one's slipping capacity to keep manifesting ones gifts. What one previously accomplished in a mundane capacity takes much more effort.
The Manifest Journey: Stage 13
Initially somewhat shaky and unfocused – the mutable. This becomes a satisfied enjoyment of life. Life is good. Creative ideas, enjoyment of sport and clarity of thought are all present. Confidence that motivates growth, and expects success.

HOLMIUM Ho

Chemical Data: Atomic No – 67,
Atomic weight 164.9,
Ring: 6 – Cosmic Spirit
Planet: Mars 2 / Scorpio (a) has been
described as a damming up sound
process, manifesting in the ordering of
substance of albumen. This is a depositing of substance, such as starch and
sugars, saved for later processes.

(b) Feeling like a mature Mars, asser-
tive, confident but with considered
actions.
Energetic Arm:
World Astral, Internal Astral
A good intuitive sense.
Physical Mode:
Fixed / Nerve Sense system
Similar elements: W

The Archetypal Journey: Stage 10
(a) fight and withdraw, obsolete, retiring confident they know better
(b) A definite sense of compromise, with repeated efforts to return to past perfor-
mance capacity; a temporary return is followed by slipping back to the state of compromise.
The Manifest Journey: Stage 12
Getting on and doing things with a positive enthusiasm. Attending to necessary
details, and routine jobs that make things work. Expanding, with a willingness to
take some risks, for the benefit of long term goals. Clear headed, with increased ESP.
Things are in order at the end of the day.

Sun 2 – Accepting Life's Reflections

ERBIUM   Er

Chemical Data: Atomic No – 68,
Atomic weight 167.2
Ring: 6 – Cosmic Spirit
Planet: Sun 2

Energetic Arm: World Astral
Similar elements: Ta, Bi
Physical Mode: All

The Archetypal Journey: Stage 11
(a) Powerless, the fight is over, weak drained empty, indifferent masked resignation
(b) A feeling of emptiness, going through the motions, a hopelessness that one will
ever return to a state in which expression of one gifts might b possible.
The Manifest Journey: Stage 11
Like its brother Pm, this element is difficult to clearly define. There is a centredness
that one can sit in, however around it are considerations of what has gone on and
how it can proceed. This can be seen as the 42 year old crisis – mid lif – where a crisis
of meaning forces us to question, why we are doing this and do we wish to continue.
There is a maturity here, but I felt impatient with the worldly affairs of the media.
The horizontal plane, (Er and Pm) has a different quality to the other elements. They
are not 'active' in any direction, as the others. They feel like they are in the middle of
things, receptive, considering and adjusting. Matter settles into form on horizontal

planes, and these elements provide the experience of how this occurs. Look, consider, question and adjust.

The Physical Group – Manifesting Form in the World

THULIUM Tm

Chemical Data: Atomic No – 69, Atomic weight 168.9,
Ring: 6 – Cosmic Spirit
Planet: Venus 2 (a) Excretion. This process takes hold of all that comes into existence as a hardening substances, due to a congestion of life processes, So all that falls out of life, and excretes it. The action of the kidneys or cellulose formation in trees.
Energetic Arm:
World Astral, Internal Physical
Similar elements: Hf

Physical Mode: Cardinal, Metabolic system
The Archetypal Journey: Stage 12
(a) Darkest black hole remedy, destruction, death, its over
(b) Sentimental about one's past; clinging to memories of a time when circumstances might have allowed for one's greater influence.
The Manifest Journey: Stage 10
A calm practicality sees you getting on with the realities of the job /task. Feeling confident with the process / product and looking for ways, and associations to expand with. This Venus is much 'harder' than Venus 1. The world has bought some reality, and 'battle hardness'. The Venus 2 process of excretion, suggests that once one can be responsible for cleaning up the dross, to stem decay, a stage of sustainability has been reached.

YTTERBIUM Yb

Chemical Data: Atomic No – 70, Atomic weight 173.0,
Ring: 6 – Cosmic Spirit
Planet: Mercury 2 / Virgo (a) Unfolding its formative forces through flowing movement. Thus space is created which is dead and falling out of lfie. In this way form can be created which serves as supporting organs. eg wood formation out of the living cambium
Energetic Arm: Internal Physical
Physical Mode: Mutable, Rhythmic system
Similar elements: Pb

The Archetypal Journey: Stage 13 (a) Decay, over, rags, philosophy, inventor
(b) The dream of potential influence exists only in theory – 'if things had been different, those gifts could have changed the world'.
The Manifest Journey: Stage 9
The Virgo qualities of sifting through the details for what is important to get on with first. A desire to run errands and do what needs to be done, to move things along. A calm practicality, with a sense of gaining momentum.

LUTETIUM Lu

Chemical Data: Atomic No – 71, Atomic weight 175.0,
Ring: 6 – Cosmic Spirit
Planet: Moon 2 / Leo (a) the sequence of generations continues through time. In a process of reflection the past becomes an image in consciousness.

There is a congestion of the life forces so a rhythmic division of cells, leaves and flowers occurs in the growing plant.
Energetic Arm: World Physical
Physical Mode: Fixed / Nerve Sense system
Similar elements: La

The Archetypal Journey: Stage 14
(a) Play, humor, finished, over, abandoned, exile.
(b) All hope for potential influence has died, but nevertheless residual ' last stand resistance' to accepting the present reality remains.

The Manifest Journey: Stage 8
In keeping with the Moon theme, this is a very gentle remedy. The actual game has begun but still baby steps. This is the first few weeks of a baby's life, when it feels like a little grub coming to terms with its newly freed body. There is an observation of the real field of play, and lists being made of what needs doing, but little Will to take the first steps. There is no rush. Thinking was clear but focused on what is right in front of you. No sense of defeat or loss, just a cautious movement forward.

Many thanks to Martin Grafton for sharing his valence and Scholten research on these elements.

Other contributions:

Planet: Moon 2 / Leo
(a) Dr Lievegoed
(b) Atkinson

The Archetypal Journey:
(a) Rochelle Marsden – Homeopathy World Community
(b) David A Johnson (Hpathy.com)

The Manifest Journey: Atkinson

Planets, Zodiac and The Chemical Elements

Throughout this text there has been reference and diagrams, outlining the double planetary rulers of the elements. Along with this reference comes a question of what zodiacal relationships could be made to the Periodic Table. Seeing as though it is now a circle it is reasonable to assume there would be a correspondent reference for the Zodiac.

My approach to this question extends from a similar question, that had to be asked of Dr Lievegoed's work on the plant processes relationship, to planetary influence. While he gave a wonderful image of this relationship, he made no mention of the Zodiac. He gave images of the two sided planetary activity, and gave similar images of the biodynamic preparations relating to these processes, however he did not give any indication for how we were to specialise the preparations, to take advantage of either of these individual activities. It seemed that if the zodiac relationships could be established, then refining the tasks of making, and application of the preparations, could be carried out during periods when these constellations were activated. Particularly when the ruling planet was within its ruling constellation, or when the Sun or Moon's transits of these constellations occur.

Lievegoed's book is a difficult read, mostly due to it being the text of a talk he gave, and thus it is quite minimal in its size, but also it takes certain liberties with the basic knowledge of the reader. Thus some of the images and concepts seem a bit foreign at first. I read it probably 20 times before I really started to feel I had an inkling of what he was saying. Sometime in the early 2000s, Dave Robison from Oregon edited the original text, into a more acceptable form of English, which helps tremendously in a basic comprehension of these ideas. So when I came to look at finding the zodiac references I used his text as the basis.

There is another book, within the biodynamic lexicon, called "Nature of Substance" by Dr Hauschka. Here he outlines an image of chemistry based upon the 12 'dominant' elements, found within the three great spheres of our environment. In the atmosphere we find Hydrogen, Nitrogen, Oxygen and Carbon. In the oceans we find Sodium, Magnesium Chlorine and Sulphur, while in the Earth /Geosphere we find Calcium, Silica, Aluminium and Phosphorus. He gives wonderful images of how these work together in their spheres to anchor various activities, which in each case he relates to a constellation.

Sadly he uses the seasonal references of the northern hemisphere, to support his argument. Being a southern hemisphere resident, naturally these references are not compatible with my experience of these constellations, and initially I did have 'a reaction' to his relationships. Nevertheless, as a Steiner chemist he is using the same reference system as Lievegoed for most of his images, so it is possible to play them off each other to see where similarities could be established. To add to this, I have my own 45 years of experience of the zodiac to add to the mix. To make things easier still is the fact that the choice of which zodiac fitted with which planet, was really only a choose between two possibilities. So if something really fitted with one reference then it stood that the other relationship must be so, whether it appeared to fit or not. So overall while it took sometime, it was a 'doable' task. Eventually I published my suggestions as part of 'Energetic Activities' (16), as I felt it was a natural addition to RS suggestions for the planets within the Agriculture Course. RS hinted at the planetary relationships to the dual Cosmic and Earthly processes, but did not say much more. So Lievegoed's efforts were a tremendous addition, while I hope my efforts have helped this study along somewhat as well. If you are interested in pursuing this topic further, Enzo Nastati has extended this topic, through providing images of how these planetary activities will act, when interacting with the 4 energetic activities. (15)

As a reference to Hauschka with Lievegoed, I can provide the following diagram, while for the whole Periodic Table see the next page.

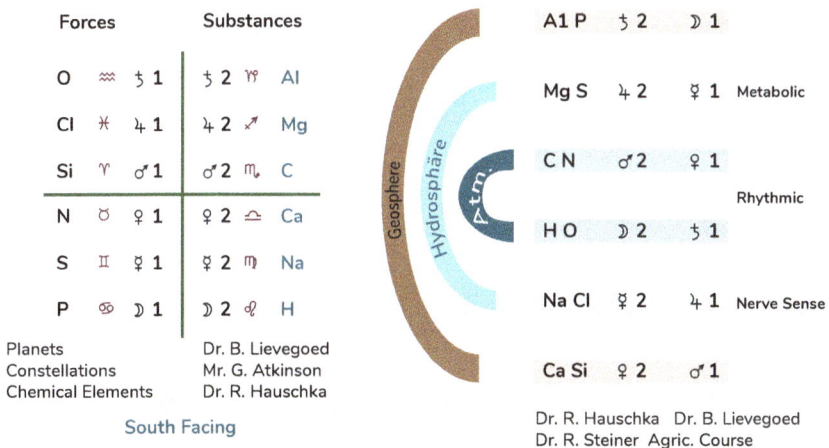

Forces			Substances		
O	≈	♄ 1	♄ 2	♑	Al
Cl	♓	♃ 1	♃ 2	♐	Mg
Si	♈	♂ 1	♂ 2	♏	C
N	♉	♀ 1	♀ 2	♎	Ca
S	♊	☿ 1	☿ 2	♍	Na
P	♋	☽ 1	☽ 2	♌	H

Planets — Dr. B. Lievegoed
Constellations — Mr. G. Atkinson
Chemical Elements — Dr. R. Hauschka

South Facing

A1 P	♄ 2	☽ 1		
Mg S	♃ 2	☿ 1	Metabolic	
C N	♂ 2	♀ 1		
			Rhythmic	
H O	☽ 2	♄ 1		
Na Cl	☿ 2	♃ 1	Nerve Sense	
Ca Si	♀ 2	♂ 1		

Dr. R. Hauschka   Dr. B. Lievegoed
Dr. R. Steiner   Agric. Course

It may be noticed that Hauschka and the Gyroscopic Periodic Table place the planets and zodiac in opposite relationship to the Anions and Cations. They are two different systems. Hauschka uses only 12 elements while the periodic Table has 120. Each system will have its truth within its context. Hauschka's combinations of elements show up very clearly in 'Alchemical Chemistry' (see p. 155).

This is a Thought Form. Information is collected, it is organised, combined and projected. It can then be explored and experimented with.

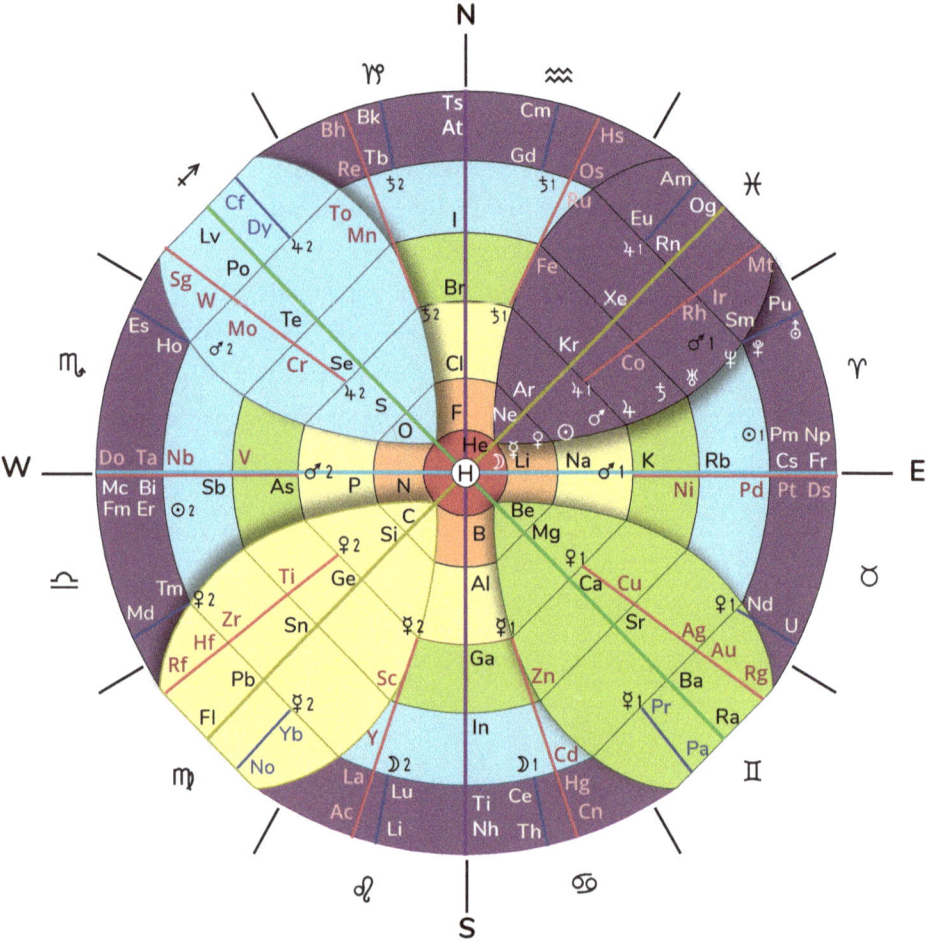

**Glenological Chemistry,
Planets and Constellations**

Magnetic North Orientation

3D Gyroscopic Periodic Table

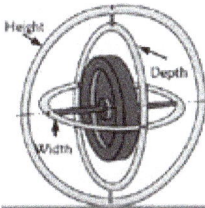

So far I have presented all the diagrams as 2 dimensional images, when in fact they need to be seen as 3 dimensional, given chemical elements are three dimensional spherical beings, and harmonic with the large 3D structures of creation like the Galaxy and Solar System etc.

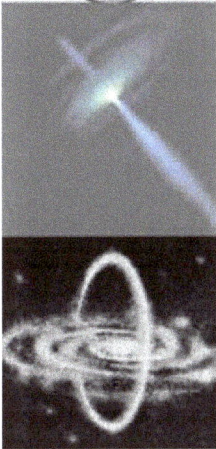

In "The Overview", I presented the two main reference structures of Biodynamics. These being the 'rings' gyroscopic diagram of the 'Cosmic' spheres of the Galaxy, Solar System, Atmosphere etc, and the 'moving vortex' gyroscopic diagram, of how the energetic bodies; Spirit, Astral, Etheric and Physical bodies, organise themselves within living organisms. So each of these images have to be moved to 3D images.

Classically, the gyroscope can be described as having three main axis. Height, Width and Depth. Two of these, height and depth are on the vertical axis and are at a 90 degree angle to each other, while width is on the horizontal.

Within the biodynamic model we identify the vertical plane(s), as associated with the Silica and Spirit / Force processes, while we associate the horizontal plane with the Calcium and Earthly / Substance processes.

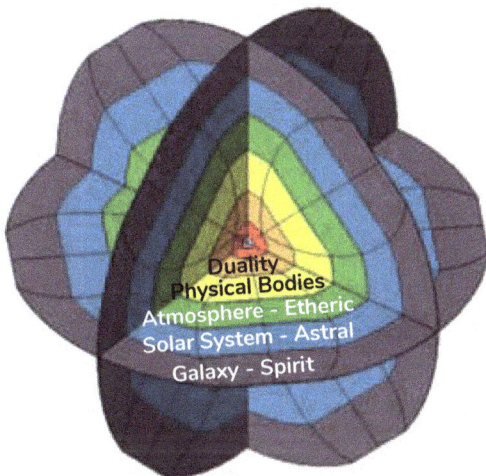

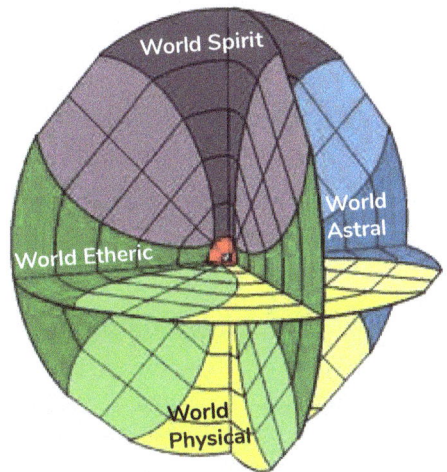

In Astronomy, we identify the horizontal plane in Galaxies and Solar Systems, as the plane of the matter buildup. This is the flat plane of where the stars and planets are in their respective structures. The vertical axis is often invisible in astronomy, however it is the plane where the vertical vortexes draw forces and matter into the center of the 'organism', before it is squirted out along the horizontal plane.

The third axis, the second vertical axis, is not commonly seen within galaxies or Solar Systems, however recent photographs of 'Gyroscopic Galaxies', and diagrams of the Oort Field, provide images of the usual flat plane of the spiral galaxy, with a perpendicular ring around it.

This is a very interesting form, that will be of special interest when we come to observe the 3D organisation of the Periodic Table. For now though it is enough to observe this as the second vertical axis.

When we draw the 'Cosmic' ring spheres of our environment as a 3D picture we obtain the image (a). Naturally the spheres within this diagram should be circular. I have left the straight lines in these diagrams to provide continuity with the earlier diagrams.

Likewise the same process can be done for the Vortex gyroscope (b) – where the energetic bodies organise themselves, once movement is bought into play – we have this second organisation on the bottom right picture.

This 3D organisation of the Vortex gyroscope has bought about an interesting insight, that is not immediately obvious when viewing the 2D diagrams. In the 2D diagram I presented, on page 38, the primary vertical axis of the gyroscope, are where the World Physical and the World Spirit were placed. The World Etheric and World Astral are the opposite poles of the horizontal plane. This follows indications from Dr Steiner given in "The Anthroposophical Approach to Medicine" (1922). In my applications of this 2D diagram, especially when applied to the Periodic Table, it became clear, that when viewing the diagram from a northern hemisphere Sun orientated perspective, the World Etheric would be on the left hand side of this diagram and the World Astral would be on the right hand side of the diagram. This keeps these diagrams consistent with the dominant and common way of orientating birth charts, as well as many other octagon cultures orientation. See page 145 for a discussion on orientating things from magnetic north, instead of facing south towards the Sun.

When moving to the 3D, I have taken it that each of these three planes, has the basic fourfold organisational pattern at their base. So when these individual planes are put into three dimensions we see the four primary vortexes, are maintained as full vortexes, of their particular body, ie a Spirit vortex, an Astral vortex, a Etheric vortex and a Physical vortex, as is suggested in the 2D diagram. However, something else appears, **when the second vertical axis crosses the horizontal axis**. Here we have a region where the Physical zone of the horizontal axis (yellow) crosses the Etheric zone (green) of the vertical axis, at the Nadir point. Likewise, on the other side of the 3D diagram, we have the Astral zone of the vertical intersecting with the Spirit zone of the horizontal axis, at the Zenith. This order mirrors a significant reference in life. These bodies work together in these partnerships, externally in nature, as Water and Earth processes moving upwards from the Earth, meeting the Light and Warmth of the Atmosphere. Similarly we find this polarity in the Human with the Spirit and Astral coming from the nerve sense region, while the Physical and etheric come from the Metabolic system.

It is at the point of meeting of these four elements at the oceans surface, that Blue Green Algae 'emerged', sparking evolution. Similarly the bodies these elements carry work this way in the various kingdoms of nature, giving us all the manifestation we see.

A more in depth exploration of Dr Steiner's use of the 3D gyroscope is presented in Appendix 2 on page 214. Interestingly he orientates his story about lifes manifestations, off the Zenith rather than Magnetic North.

The 3D Periodic Table

Within the process of 'observing' the Periodic table, there are three groupings of elements, which each vibrate at a different frequency. This suggests they each inhabit different harmonic range of space. These three dimensions are, the 8 Major element Arms that work on the fourth harmonic, the 10 Trace elements that work on the fifth harmonic, and the 14 Actinides and Lanthanides which work upon the seventh harmonic. Each of these have some specific characteristics. In my earlier diagrams I placed them on top of each other to identify their interrelationships, however they should be seen as individual activities, within the greater whole.

The 'Major' 4th harmonic elements form the basis of manifestation. They are the base elements that provided the structures of many forms and activities. They are found at every ring of the Periodic Table. Generally, the elements in the inner rings are those that most actively support life, while the further out in the spheres one goes, the more toxic the elements become. Very few of the major elements found to support life are placed beyond the fourth or Etheric (green) sphere of the Periodic table.

The major elements begin with Hydrogen, which is placed at the very center, of the Periodic Table. This central position, when found in the Galaxy or Solar System or the human, is the position of the internalised spirit. It is the central core of 'the individuality', at whatever level. This provides the focusing center, or spirit, around which everything in that dimension organises itself. In the 3D

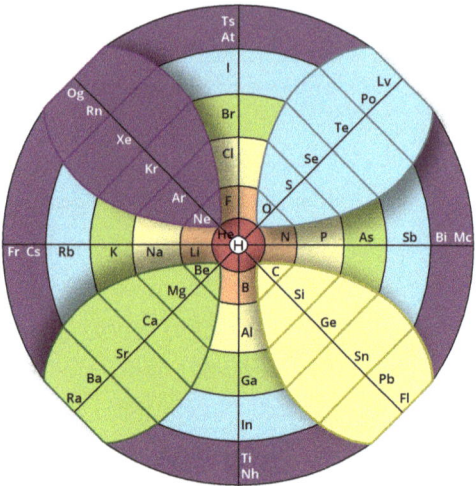

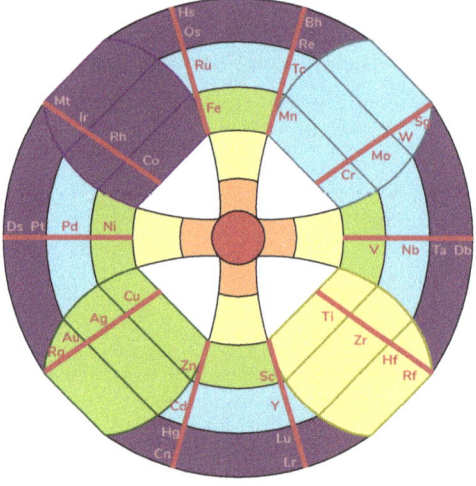

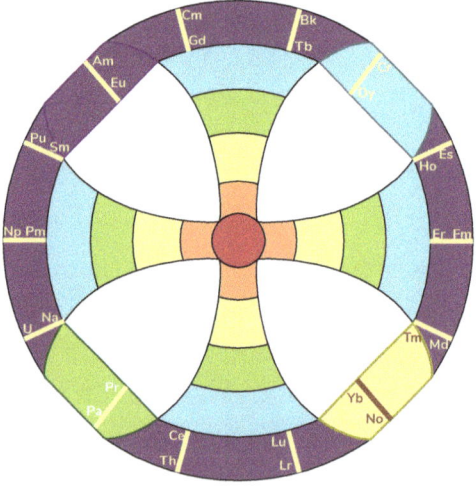

form, these Major elements – the fourth harmonic structural elements – would be placed on the fourfold primary vertical plane of the gyroscope.

The Trace, 5th harmonic, elements are those that become active as catalysts in biological processes, and allow life processes to come into formation and be maintained. These do not start to appear on the Periodic Table until the fourth /etheric /life body sphere. Lifeforms can only begin to form when the etheric body becomes active in creation. It is this life element in all things and its incarnation and exit, (along with the spirit), that is the key element in the difference between life and death. Within the organised gyroscopic spheres of our environment, the matter always appears along the horizontal plane. Thus life within the gyroscope, is found on the horizontal plane, and therefore the transition elements are best placed on the horizontal plane.

The Actinides and Lanthanides are rare earth and radioactive elements and are found on the outer most rings of the Periodic Table. This outer ring, being the galaxy, is the Cosmic Spirit sphere. If the center of these diagrams represent, the Unity, and the placement of the Individualised spirit, this outer ring is the Cosmic Spirit.

Within the biodynamic reference of plant growth we are presented with two Silica processes. One coming from the Earth, up through the inside the plant moving upwards, called the Cosmic Silica / Cosmic Forces process, and another coming from the environment in the light and the warmth called the Terrestrial Silica / Cosmic Matter. This external light and warmth process strongly influences the skin formation of the fruit, while the internal silica process is found in the formation of the seed within the fruit. This outer Galactic ring of the gyroscope is the 'skin', boundary setting influence, of the Cosmic Spirit sphere.

Given that this outer sphere is a spiritual sphere, we would expect to place these 7th harmonic elements, on a vertical axis.

The next step is to put these three planes together. The picture on the next page is the result.

3D Periodic Table
Arms   Stage 2

Two Orientations

On page 31, we found there were the two orientations of the gyroscope. The axis based upon **Magnetic North and the axis based upon the Zenith**. While this difference may not seem of particular significance for dwellers of the southern hemisphere, as both are in basically the same direction and we naturally orientate to 'the north', it is a significantly different thing for the people of the northern hemisphere. They naturally orientate to the Sun, and thus to the Zenith axis, which means facing 'everything' to the South. So to orientate anything off magnetic north means they must turn in the opposite direction.

My world view has developed from my studies of Astrology, which in turn is developed from the long cultural tradition of orientation to the Sun and its path of movement in front of the Zodiac, from a northern hemisphere perspective. Rightly so, I might hear you say, however with this acceptance comes a particular orientation towards the world, that has its own peculiar difficulties.

When I began developing circular images from my 'biodynamic' vortex diagrams, (which is only one small part of the spherical gyroscope), I was naturally drawn to the frame of reference presented in the Astrological birth chart. There is several millennia of reference material developed around Astrology, so there is plenty of information to draw upon, to find the inner nature of the circle, and then to some degree also the gyroscope.

The birth chart commonly used today is however a very specific beast. I have written about its peculiarities at http://old.garudabd.org/books/4_6.html and in 'BD questions Astrological Answers'. There I emphasised how it is somewhat removed from the astronomical reality. The specific issues that concern us here is that the 'birth chart' is a map of **where the planets are placed at the moment of birth, in relationship to the background of the Zodiac, as seen from the Earth**.

When looking at the birth chart for indications of the activity of various 'spaces', it is useful to look to where the eastern horizon is at birth. This is determined from the birth time at a specific place. This allows us to mark off the 'houses'. These areas begin at the eastern horizon, and are marked off in an anti clockwise direction. As they are based upon the Earth related eastern horizon, the houses indicate the various physical areas of our life, that any particular planetary energy will emphasis, e.g. personal finances, home, work, societal interaction etc.

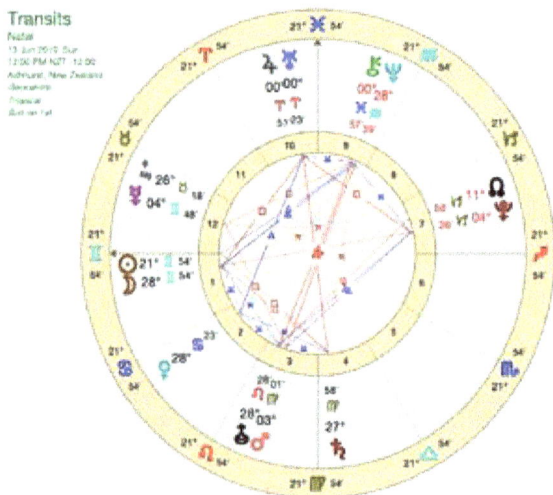

By marking off the path of the planets and the Sun, along the ecliptic, we observe the path the planets take through the night, at the 'bottom' of the chart, and where they travel through the day time being the 'top' of the chart. Where the Sun reaches the highest point on the sky during the day, we call this point the Zenith. With this being the 'highest' point of the Sun's travel, we associate this with the area where we impact the most upon the world, Symbolically, this means that the Sun, which is the macrocosmic representative of our internalized spirit, is reaching it highest expression. Thus this is the place of an individual's direction, purpose and influence, which is often ones career. It can also show as the husband's / partners career.

Given that we have marked off the Eastern horizon, we have also marked the western horizon. From this we gain the image that the Sun's zenith, would be a 'northish' point, while the night time point would be a 'southish' point. This is particularly easy to make this association when living in the Southern hemisphere as we look to the north whenever we look to the ecliptic.

However it is all somewhat more complicated than this. Astrology has developed over its 5000 year history in the northern hemisphere. In the northern hemisphere, the ecliptic, or the Sun's path, crosses in the southern part of the sky. Thus to read a birth chart, or work with any zodiac orientated perspective, generated in the universally accepted north hemisphere orientation, we must first face to the south. Then we have the eastern horizon (Ascendant) on our left and the western horizon (Descendant) on the right hand side. So this 'northish' point is actually pointing to the south.

So to read this chart correctly you must first face to the South pole.

Birth charts are two dimensional diagrams drawn on paper. Therefore we do not fully experience the three dimensional reality of the gyroscopic solar system. We do not see the different distances the planets are from the Sun, nor do we see the different inclinations from the ecliptic that each planet has. So we do not have to really confront the reality of what the birth chart is. In fact, we take a tremendous amount for granted, when we accept the western birth chart, as the 'normal' point of reference. I am not saying it is wrong, just that it is a very specific document that gives us very specific information.

So how does the birth chart relate to the gyroscopic sphere?

Within a gyroscopic sphere we have three planes, we have the circle of height, the circle of depth, and the circle of width. Thus we have two vertical planes and one horizontal plane. The vertical planes are orientated off the magnetic north point, which we call North. Therefore we have South opposite. If we stall the gyroscopes spinning movement for a moment, we can also identify a East point and a West point at ninety degrees to the vertical axis. The third plane is across the horizontal plane. So while this horizontal plane can have a east and west point, it can never face North or South, as it is always 90 degree to the axis that defines North.

This reality becomes significant when we realise that our planets are all manifestations of the horizontal plane of the Sun's gyroscopic nature.

Gyroscopic physics states that as the 'Sun' gyroscope spins, it sucks in matter and forces in its north and south poles, and then squirts matter out along its equator. The Sun's magnetic fields then organise this substance into rings, which over time have organised themselves into gyroscopic balls, we call planets. These in turn suck in cosmic dust and any other Sun 'garbage' through their north and south poles, consolidating this into its planetary substance. Thus we have the planets, lined out along the horizontal plane.

In a perfect solar system, all of the planets would move along one horizon line, however they do not. Mostly though they do move within 5 degrees either side of the horizontal. One significant thing we need to consider is the

Earth being 23 degrees tilted off its true North South axis. So in our daily orbit, we see the Suns plane move higher and lower in the sky, while to a greater extent throughout the year, this angle provides us with our seasons.

This brings us to the Zodiac. The Zodiac is that band of stars, in front of which the Sun and the planets moves. It is the band of stars wrapped around the horizontal plane of the Sun's or the Earth, gyroscope, depending on which orientation –Heliocentric or Geocentric – you choose to use.

So while we identify the Sun's midday point as the highest part of the Sun' path, and identify its 'northish' qualities, it is not really true. Firstly in the Northern hemisphere you will be looking South, and even in the Southern hemisphere, the north pole is still 90 degrees away from the point, we are looking at when we see the Zenith.

A secondary problem arises for the northern hemisphere as. When we look south to see the Sun, this places the east on the left, and west on the right hand. From this orientation, we then develop a reference system based largely upon the astronomical and natural observations. We extrapolate that the day-time (above) of the chart indicates an extrovert personality, when a predominate number of planets are placed there. While the night time area, with planetary predominance, suggests a introverted orientation. We then follow onto the quadrants of the circle, being divided into collective (left) and personal (right) areas of life, which leads us to all the house definitions, and so on. This is all very well and many fine things have been developed from this reference, over the last few thousand years. The key thought though is, this orientation is towards the Sun's horizontal plane.

This is all very well, however there is another very worthwhile orientation to consider when living as a being upon planet Earth, and this is to reference ourselves off True Magnetic North.

As electro magnetic beings, we have a natural north south polarisation within our magnetic field – head to the north, feet to the south – and some say that being aligned to the Earth's magnetic field, especially during sleep, is a beneficial experience, which helps to realign the static electricity within our bodies, back into a naturally harmonic resonance.

When we look into the Earth sculptures of ancient cultures, foremost of which would have to be the gothic cathedrals, we find they are aligned through the

North / South, East / West axis, with an emphasis of their main hall often, along the East West axis. More ancient cultures, like the American Indians, with their medicine wheel, the Buddhists mandalas and the Mayan Calendar also orientate themselves off the Earth's North South axis. This makes very good sense, as we are primarily Earth beings and therefore why not orientate off our environment. However there is still some confusion in all these structures. Many place a dominant quality on the South Zenith Summer reference, which thus diminishes the role of the north pole. Thus the seasonal reference is usually dominate in all these cultures.

How did we get from facing North to facing South? Of course the Sun's warmth caught our attention, but it appears not really until late in the Egyptian period.

When we orient off the North pole, it does not matter where you are on the planet, you will always look to the north, and therefore East will be on your right hand and west will be on your left hand. While this is not a problem for the people living in the southern hemisphere, it is a problem for the people living in the northern hemisphere, as they have to turn around and look to the cold north, instead of the warm south. It also messes up the orientation of the northern hemisphere orientated birth charts and all that is deduced from it. Not in the value and meaning of the information, but in the orientation of that information primarily to the switching of the left and right hands, especially when we place emphasis on the left being intuitive and the right to our rational conscious qualities.

This orientation issue was not a problem, while I was developing my diagrams within the one particular reference – northern hemisphere. However once I started to find this same gyroscopic reference showing up in the artwork of past cultures and in Earth monuments, such as cathedrals and especially in the Chartres Labyrinth, then this issue of how to place the astrological referenced information with the Earth orientated structures, became a problem.

In clarifying this difference I am not making one orientation right and the other wrong, I am just clarifying they are different. The significance of the difference will be investigated further, with some reference to Dr Steiner's comments on this matter (see Appendix 2 – 'The Three Dimensions of Space').

My diagrams

The implications for my diagrams however is relatively significant. My present interest (2010) is to apply the magnetic organisation of the gyroscope to the Earth and to the existing 'earth monuments' we find in things like the Celtic stone circles, the Chartres Labyrinth, the Gothic Chapter Houses, and many of the other Templar structures dotted about Europe. These 'forms of earth power' can be easily created anywhere, with any circle or octagonal form being a manifester of these organisational forces.

To date and throughout this book, my diagrams are drawn with the northern hemisphere reference. This means the next step is to change many of my diagrams, from the orientation of the Zenith to the orientation of North. Thus when you face to the north, the 'old' diagrams need to be flipped horizontally, which means they can not be seen, or you have to have them over your head. The North South pole stays the same. East and West switch over. I began this process from the 21 March 2010. My earlier diagrams are not wrong, they are just a little difficult to use, when applying them to the Earth sculptures. The information that has been generated from them is still relevant in itself, but we need to be conscious, it is developed from the ecliptic as seen from the northern hemisphere.

The journey I have taken, from the Astrological to the Earth based reference systems is in the opposite direction to that walked by humanity throughout its history. Humanity first came to the northern orientation, with reference to many different stars in the heavens (3000 BC) It was much later, (200 AD) that 'we' moved to a Sun related orientation. Various cultures, orientated initially off the Pleiades, Sirius, Vega, Canopus, and so on, all stars NOT in the zodiac.

The story of the Egyptian pharaoh Akhenaten indicates the difficulties this change to the Sun created. He was "a Pharaoh of the Eighteenth dynasty of Egypt, who ruled for 17 years and died in 1336 BC or 1334 BC. He is especially noted for abandoning traditional Egyptian polytheism and introducing worship centered on the Aten, which is sometimes described as monotheistic or henotheistic. An early inscription likens him to the sun as compared to stars, and later official language avoids calling the Aten a god, giving the solar deity a status above mere gods." – Wikipedia. Upon his son's death many references to Akhenaten were removed from the hieroglyphic record, and polytheism was re-instituted in Egypt.

What does Humanities Change of Orientation Mean

With reference to several of my earlier articles, (10) I draw your attention to the interesting process of duality we find all around us. Wherever we find two things set against each other, Tropical Signs – Sidereal Constellations, Geocentric – Heliocentric, The Cancer / Leo Zodiac as opposed to the Aries / Pisces zodiac and so on. These 'differences' mirror the duality of Spirit and Matter, or Force and Substance, Cosmic and Earthly, External or Internal organisations, as expressed in the dual, magnetic and seasonal crosses. The conclusion of these explorations is, the more we work with what is astronomically real, the closer we are to the archetype or primal vibration, and thus the closer we are to spirit, while the further we move away from the real astronomical phenomena and move towards something concocted by Human perception, such as the Signs of the zodiac, the closer we move to Matter, or some abstract aspect of matter. In the case of the constellations versus the Signs of the Zodiac, the archetypal information, of the constellations, will describe very unconscious and even collective aspects of the human, e.g. our biological evolution and function, while the more abstract reference, e.g. the signs, in particular, will talk not so much to our body's organisation, but more to the human's psychological organisation, which is a very abstract thing, in the context of the physical universe.

With the two orientations of this article, we have one focused upon the magnetic northern pole of the Earth, and the other is focused upon the apparent path of the Sun around the Earth, and the small ring of stars behind it. The key feature of this change, is our move to a purely Earth related phenomena from one based upon the Sun and the zodiacal stars. Thus there is a move from our single Star, the Sun, which is the individualising aspect of World Spirit sphere and therefore metaphorically relates to our internalised individualised spirit, to something more physical, yet something very astronomically real.

Initially humanity focused upon the stars in a general manner, which showed their relationship to the Cosmic and collective World Spirit, which changed with the process of humanity focusing upon our Sun as THE deity. This change reaches a peak when we moved from polytheistic religion to monotheistic religion, with the single God, being considered a Earthly representative of the Sun being. While Akhenaten had a go at this, (Judaism has non solar monotheistic tendencies) it was not until Christ's incarnation, and his reference as a solar deity, that Solar monotheism really took hold.

The evolutionary significance of this change, is the reorientation of our focus away from being humans at the mercy of the Gods, (Greek polytheism) to us as individuals, firstly in direct relationship with THE God, (post Luther Christianity), and then to us now as direct personal representatives of THE Solar internalised God. We have moved from the collective to the personal. In correspondence with this development, we can track that we have moved from a empathic 'clairvoyant' consciousness, to the more rational scientific and defining consciousness. Thus we have moved away from the Earth within the many starred Cosmic dome, to the Sun as its own central reference.

My chapters on the 'Birth Chart' indicate just how far we have traveled, in the process of identifying our personal consciousness' karmic life, here on Earth.

Stepping 'forward' to an astronomical Earth orientation, based on the Earth's magnetic pole, can be seen as either a further internalisation of the Spirit, thus 'individualising the Earth', and a symbol of a further internalisation of the Sun principle, and thus full spiritual responsibility for oneself or simply it is a step to more practical and physical applications, which become useful for doing stuff on Earth, like dealing with chemistry and agriculture and energy organisation on Earth.

My interest in this is two fold. Firstly, we have the energetic reality of the Earths electro magnetic field. This is a free source of energy, that is always

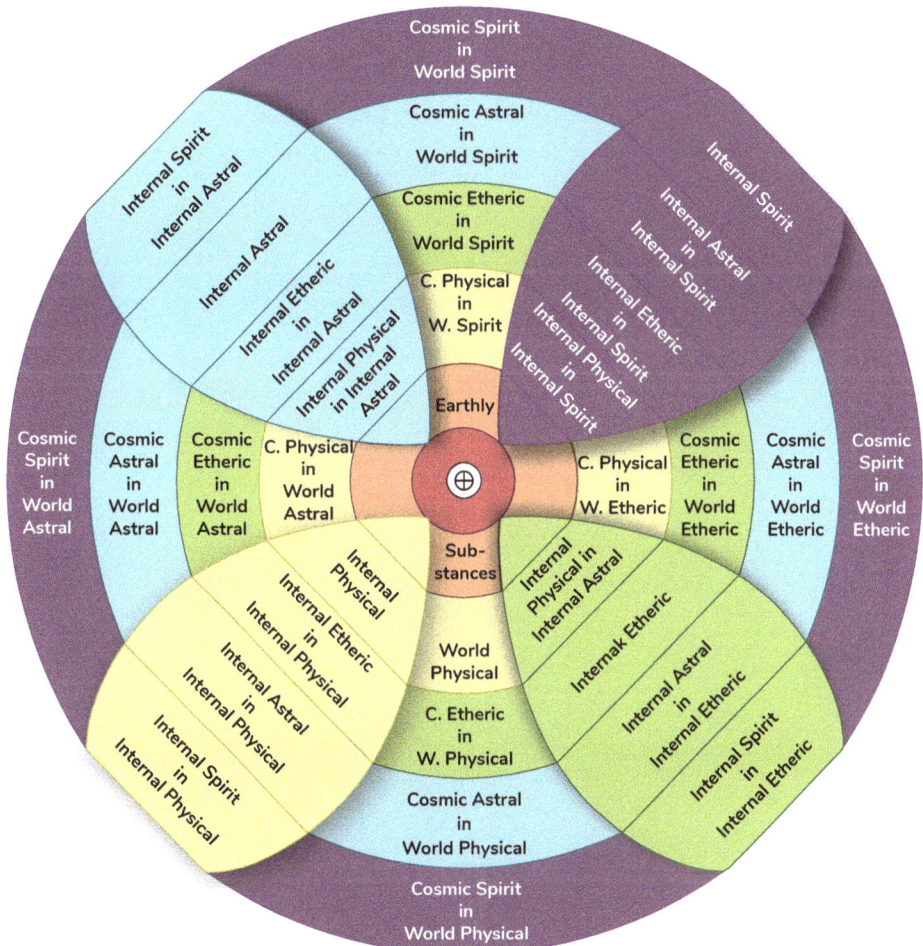

present, forever being regenerated by the mere process of the Earth moving through space. Thus this is an energy field at our disposal, 24 hours a day anywhere on the planet. Why not make use of it, as a energetic recharging source for the unconscious processes still active within our bodies, and especially for the beings still embedded fully in this matrix, the soil, plants and animals?

Secondly, because we have many Earth orientated structures, dotted about the planet created by past civilisations, who knew of this 'free' power. These ancient places are placed on points of intensified earth energy, which makes them especially worth visiting. The biggest problem though is we do not have much conscious information about what these places are, and what they can do.

The archetypal organisational reference, I have developed along the Astrological Biodynamic journey, is directly relatable to the Earth magnetic field, and therefore it provides an interpretative image of the organisation existing within these Earth structures. Thus we can have an image of the function of the complex of energies present, anytime we draw a circle or an octagon.

Orientating off Magnetic North

The diagrams presented so far in this book, all arise from the traditional northern hemisphere preference for orientating most things off the Sun, and thus one needs to stand facing the south for the East to be on the left hand. In 2010 I moved to orientating myself and thus my diagrams to magnetic north, and have presented more considerations of why, in "From Double Cross to Gyroscope", Appendix 2. This orientation arises naturally out of bringing everything as close to reality as possible. I include the previous two diagrams as a reference.

Alchemical Chemistry

The earlier chapters outline the basic structural form of the circular and then gyroscopic form of the Periodic Table. From this it was possible to identify the energetic qualities of all the elements of **the Periodic Table**, within the language and context of Dr Rudolf Steiner's (RS) world view.

There are many practical applications that can be derived from this. However after a 'wait there is more' moment in my process, it became obvious that a further relatively obvious and simple step, can open up a whole other vista for the application of the Periodic Table organisation. In following the goal of finding practical ways to apply all this to nature, the thought arose that, **if we wish to see this picture and its information, more orientated to physical life, then we could change the orientation of the gyroscope from standing upon the World Physical Arm, – where it shows us archetypal external order and functioning, – to standing upon the Internal Physical Arm, where it should show how the elements work on internal chemistry of life forms**. I first presented this image in my 'Biodynamic Agriculture' thought form, on the garudabd.org website around 2000. It is only in 2013 though, that a discussion with Mark Moodie, around how one might 'threefold the threefold', or find three aspects of each of the 'old' alchemical references, Salt, Mercury and Sulphur, that this step really came into its own. I had long been looking for 'the way' into the Periodic Table, where it all 'just makes sense'. This was the intention right from the start in 1996, however my basic knowledge was insufficient to 'see' it'.

When its time had come, the 'money shot' of this step is how this orientation relates to Dr Steiner's medical lectures, especially the 1920, 1921 series, where RS focuses his attention on the old alchemical 'threefold' reference of creations processes. As usual he gives it his own spin, which does not necessarily conform to the expectations of 'some alchemists', however his take is consistent with his view of the Agricultural Individuality', and the indications provided by the Astrological model.

Within the 'Biodynamic Model' presented in 'Biodynamic Decoded' the threefold stage has been identified as the dominant process found within the physical body. From Astrology this is the layer of the 'Modes' and characterised as the consolidating Fixed activities, the expansive Cardinal processes and 'in-between' Mutable processes that arise as a dynamic expression of the interaction of the first two processes. This same image is used by RS in his description of the physical processes, where he describes the nerve sense processes, centered in our head region as contractive, with the metabolic activity centered in our 'belly' as expansive. The Rhythmic system which includes our circulation and respiration, comprise the middle sphere, that mediates between the activity of the two poles. The middle is only healthy when the poles are working together properly.

This activity was imaged as:

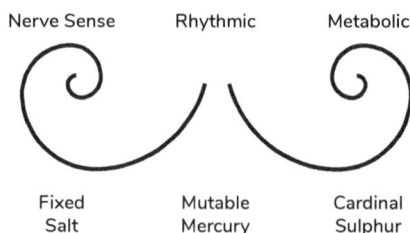

Nerve Sense   Rhythmic   Metabolic

Fixed   Mutable   Cardinal
Salt   Mercury   Sulphur

The developments shown in the following pages take this basic law one step further and looks at how each of these three building blokes of the physical form can be divided into three parts.

The outcome of this whole journey is its arrival at a very practical, safe and potentially free avenue of health care for all who can be bothered to understand it. It is precisely because of all these qualities, and that RS appears to have already made the bones of its application available in his lectures, that I feel it is necessary to make this 'coordinating' piece of information available to whoever wants it. The potential value to humanity far outweighs my need to exclusively benefit from it.

I will outline the various stages of the logical unfolding of the understanding, that leads to the final organisation. This system does not arise from faith or mysticism. It is **a rational scientific development of one fact upon another**, albeit using a reference system beyond what materialism allows itself, but nevertheless, it follows sequentially along a path that provides clarity and application to RS indications.

I thank him wholeheartedly for leaving behind what he did, as I very much doubt my efforts could have reached this point, without him. To me he is indeed 'Saint Rudolf'.

The Periodic Table of Life

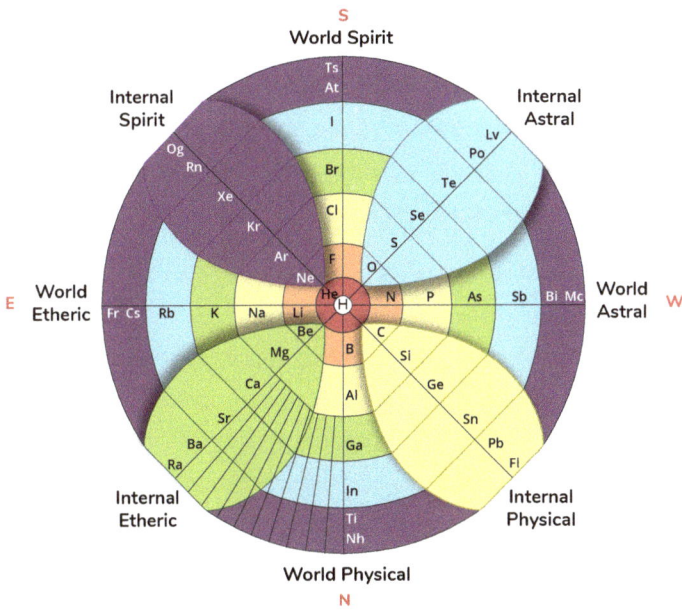

In earlier chapters the circular periodic table was shown as it stands in creation.

Firstly it was shown orientated as if we are looking, in the north hemisphere towards the Suns ecliptic. This means facing to the south, with the East on the left hand – as is the tradition of many Piscean age cultures. The key thing to notice here though is that the main vertical axis is the World Physical, World Spirit axis.

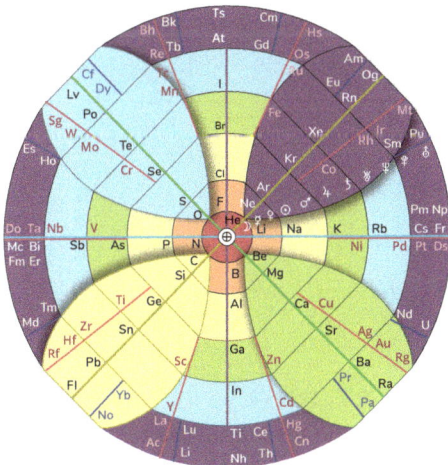

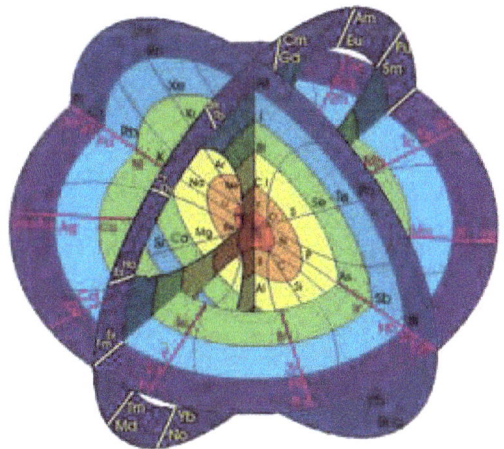

The next step that was taken, on page 157, was to change the orientation from the northern hemisphere focus upon the Sun, to focusing the same vertical axis upon Magnetic North. Chemistry, as an expression of the electro magnetic nature of creation, and thus conforms to the real electro magnetic nature of our Earth. It does not conform to our human desire to be warm or worship of a Sun god. So this change of orientation is necessary once we begin to work practically with the periodic table.

(This basic difference of orientation, was the cause of the 'disorientation' I display during the third DVD of my 2014 presentations, which are now available as 'Energetic Activities'. I am now use to working with the magnetic orientations, but still need to teach from the northern hemisphere ecliptic pictures, as these arise from all the previous stories, developed from the several thousand years of previous northern hemisphere references.)

A further step was taken when the 2D representations of the periodic table were extended into the 3D spherical form. In Appendix 2 I explore the three dimensional nature of space, and this provide references for the practical application of this 3D form. I have produced this form into a hanging sculpture, that can be used practically in ways outlined later.

So, the earlier part of this book was concerned with a description of '**what is there**'. The energetic activity of each element was identified, and some images of the elements interactions and relationships with each other is provided. Practical experiments have shown there are many applications that can be found for using the elements energetic activity to move the bodies around, in much the same way the potentised BD preps do.

The question arises though, what can be done to focus our view more sharply upon how this all effects life processes on Earth?

If we want to learn more about how the chemical elements are going to work on the internalised physical body of life forms, then we can shift the vertical orientation of the circle, from the World Physical body to the Internal Physical body. The following diagram is the result. In doing so we are changing the orientation from the primary archetypal cross, to the secondary manifest cross. Please note this picture is also orientated to magnetic north and so East is on the right hand.

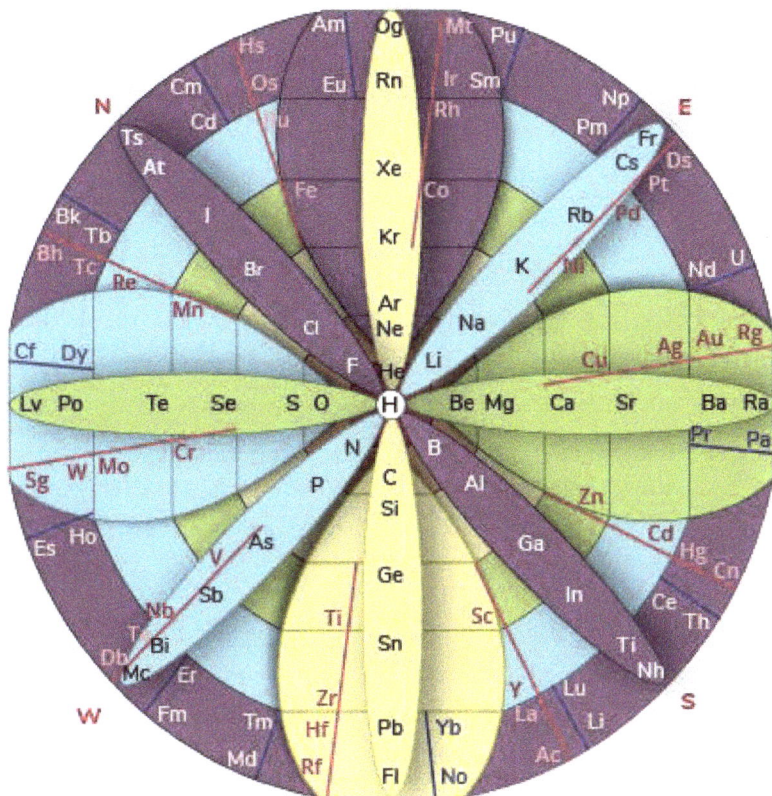

As a small aside – A close inspection of Dr Hauschka's zodiacal relationships to his 12 elements, shows his cycle of the zodiac e.g. page 155, 'Nature of Substance', are placed as if one is looking towards the north. They would rise out of the east and move to the left, rather than moving towards the right side, if one looked south. I do not remember any comments about this in his text. This suggests he was aware of the need to focus chemistry upon the magnetic reality.

Once the shift of axis is made we can begin the perceptive process of **what can be seen**. The first thing that strikes me is that at the 'base' of the physical arm is the element Carbon, followed by Silica. Carbon is the basic element of organic life. We are all carbon based life forms, while Silica is the element that forms the 'formative' scaffolding upon which the Carbon is placed. Carbon is quite a special element, in that while it exists as a gas, and in physical forms such as coal, the carbon we have are deposits from life forms. It does not readily combine with any elements other than it sisters in protein, Nitrogen,

Oxygen and Hydrogen. So it is a element intimately linked into the chaotic processes of life, and in many ways is the molding substance that becomes an expression of the environment within which it finds itself.

Opposite to the Internal Physical arm is the Internal Spirit arm of the noble gases. These gases have a complete outer ring of electrons (8), and so do not easily interact with other elements. Their motif is that of the autistic spectrum of human psychology. They are individuals who do not feel the need to socialise or interact with others, however as we know autistic people sure do impact on their environment. Thus the noble gases may not 'talk' to anyone, but they do provide 'atomic weight' to a environment, as and when they see fit. So they too suggest they can act as chaotic elements that respond to the environment they are in, and thus to the needs of the time. Therefore the vertical axis of this diagram can be looked upon as being Mutable, or responsive to its environment. So what is that environment?

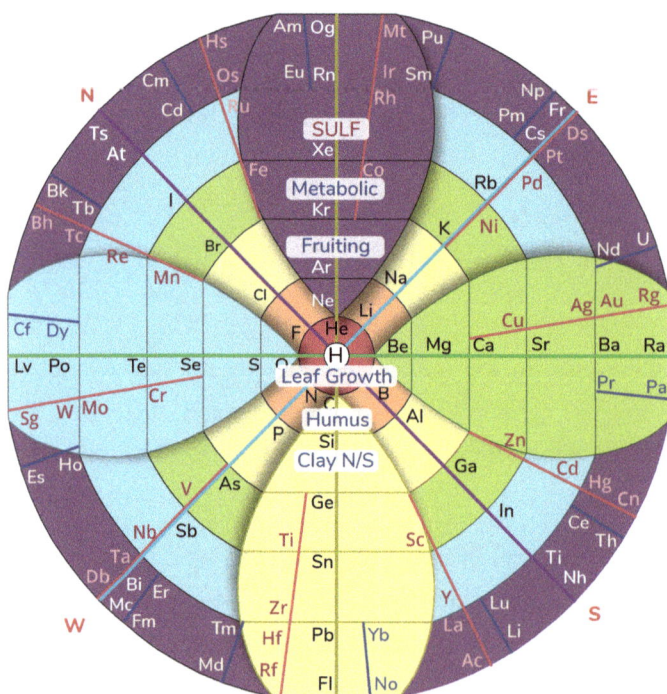

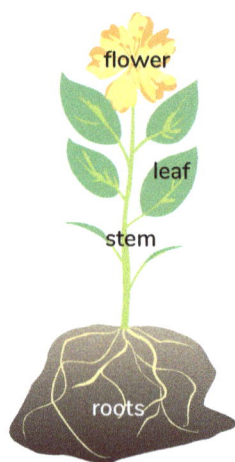

Just looking at the organised structure that is presented to us, we can identify that clay is made of Aluminum Silicate, and it readily locks up Phosphorus. Above that is a layer with Boron, Carbon and Nitrogen. These three elements are major components of Humus, the central component in the 'living' layer of soil. Humus provides us with the most stable and usable sources of both Boron and Nitrogen for agriculture.

The next layer up, has Calcium, Magnesium, Oxygen and Sulphur, all of which are central to the proper develop of leaf growth. Magnesium is the central element of photosynthesis, that occurs primarily in leaves, while calcium along with oxygen, are the carriers of the Etheric activity, and necessary to grow large and nutritious leaves for fodder. While Sulphur is necessary to activate the many biochemical processes needed for life to occur.

The next layer of Lithium and Fluorine, (which liquefies Silica), do not find much use in agriculture. However the next two layers of Sodium Chloride and Potassium Chloride do. Sodium Chloride is essential for the regulation of the digestive acid processes found in our stomachs, while Potassium Chlorides use within agriculture is for the development and sizing of fruit. RS shows, the alkalis are akin to a plant process taking place within our metabolic function.

So within the Periodic table itself, is the form of the threefold plant and human. Below the middle line is the 'nerve sense' soil, and above it is the 'metabolic' atmosphere. Therefore we can identify, the below region carries the contractive SAL qualities of the nerve sense system, while the above section of the Periodic table carries the expansive SULF qualities.

3 fold x 3 fold

These references brings us back to the imagery we looked at in the 'Overview' chapter of this book, in regards to Agriculture. We have placed the threefold qualities onto the periodic table, but is it possible to **3 fold this 3 fold**?

First let us observe that so far we have the threefold organisation setting the vertical structure, on the Silica / Argon arms, and also that the 'middle axis' of this picture is comprised of three arms. This suggests a very mutable and responsive 'being', – the vertical – awaiting to be influenced by something 'outside' itself.

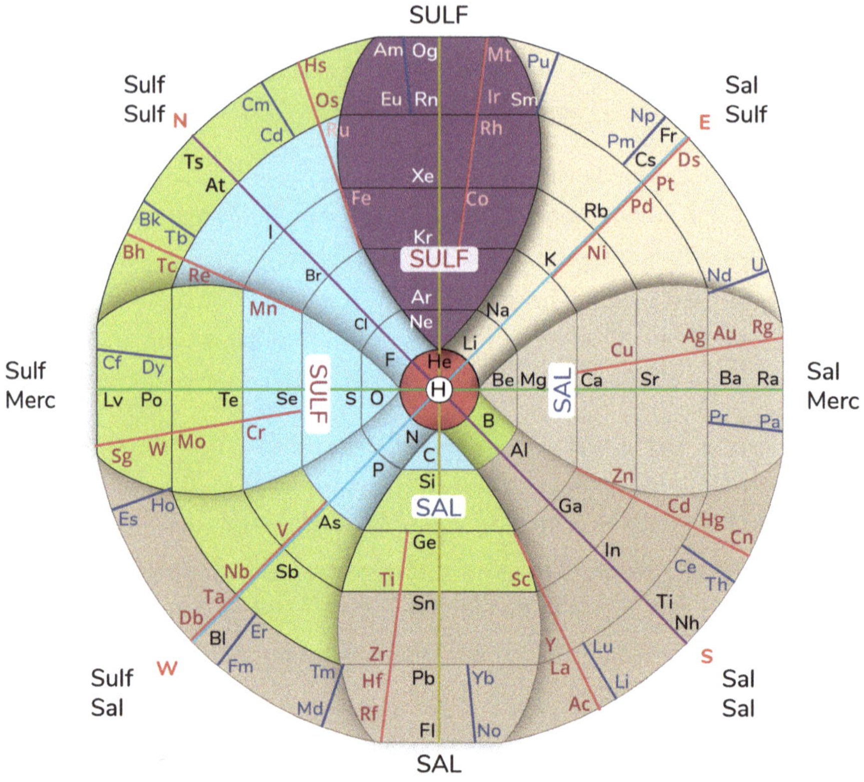

The Metallic States and Alchemy of the Chemical Elements

We are looking for practical, physical indications, so we can look to the Cosmic Physical (yellow) ring of this diagram. Here we find Si at the base and its polar mate is Argon, above. To either side of these we have two groups of three elements. On the left we have Sodium (Na), Magnesium (Mg) and Aluminum (Al), while on the other side we have the elements of Chlorine (Cl, Sulphur (S) and Phosphorus (P).

We can observe that the first group are a group of positively charged cations, while the left hand group are negatively charged anions.

Our next observation arises when we look at the diagram coloured so as to indicate **the 'metallic states' of the elements**. All the brown coloured elements are metals of various natures. The green elements are 'in-between' elements of metallic and non metallic nature, while the blue elements are all non metals. The mauve colour indicates the noble gases.

From this picture it is clear that the elements on the right are much more dense and solid, and thus associated to a SAL process. While the elements on the left are softer and more reactive, and thus more associated with the active SULF processes.

From the layout of the diagram, we can identify that the threefold Sul elements can be observed as, the P being expected to act as a Sul Sal quality, S is a Sulf Merc quality, while the Cl has a Sulf Sulf quality. Similarly the Al has a Sal Sal nature, Mg has a Sal Merc quality while Na is the Sal Sulf element.

These associations have developed out of looking at the diagram of the circular periodic table, seeing what is there, and making correspondences to information we have already accumulated throughout our biodynamic journey.

One of those pieces of information is Hauschka talking about the archetypal significance of the combined elements Aluminium Phosphate (and their partners Calcium and Silica) forming the Geosphere, while Magnesium Sulphate and Sodium Chloride are the four elements making up the Hydrosphere, while Hydrogen, Nitrogen, Oxygen and Carbon make up the Atmosphere. All of these elements have pivotal positions in this story as well. All he says about these relationship can be bought into our understanding here.

We can look further into the qualities and the relationships, that are known about these elements to bring depth, texture and context to this image. Luckily we have Dr Steiner's 1920 and 1921 medical lectures to draw upon to provide some further gravitas to this image.

I would love to provide extensive quotes from RS to illustrate his associations, however this process will have to be left to the another edition of this book. I will need to re read all these 30 or so lectures and craft a more academic presentation of this material, which will take time. At present I am wanting to present the overview approach to this study.

So in a shortened version, I can provide some of the imagery RS gives for this outline with references of where you can find it in detail.

Continuing on with the physical body elements, we can build up a picture of references. On page 164 is a diagram of some of those RS makes.

Alchemical Chemistry Picture

In 1920, RS talks of the threefold processes as Salt, Mercury and Phosphorus. Traditionally in Alchemy, Phosphorus was included in 'Sulphur'. While I have no evidence that he had the actual picture I present here, he obviously had it in some form of imagination. His Phosphorus emphasis, does accurately describe the Sulf aspects of the nerve sense activities of the head, he often referred to. Similarly from his descriptions, and from the fact that Sodium is the controller of the water and thus the Etheric activities of the organism, then we can see he is generally describing the less obvious 'extremes' of both Sulf and Sal, in his initial stories. It is important to appreciate that in the original alchemical stories, they are not talking of the actual substance Salt or Sulphur, but a process, that in the 1600s they could observe expressed most obviously in these two substances. For RS, he has chosen to focus on

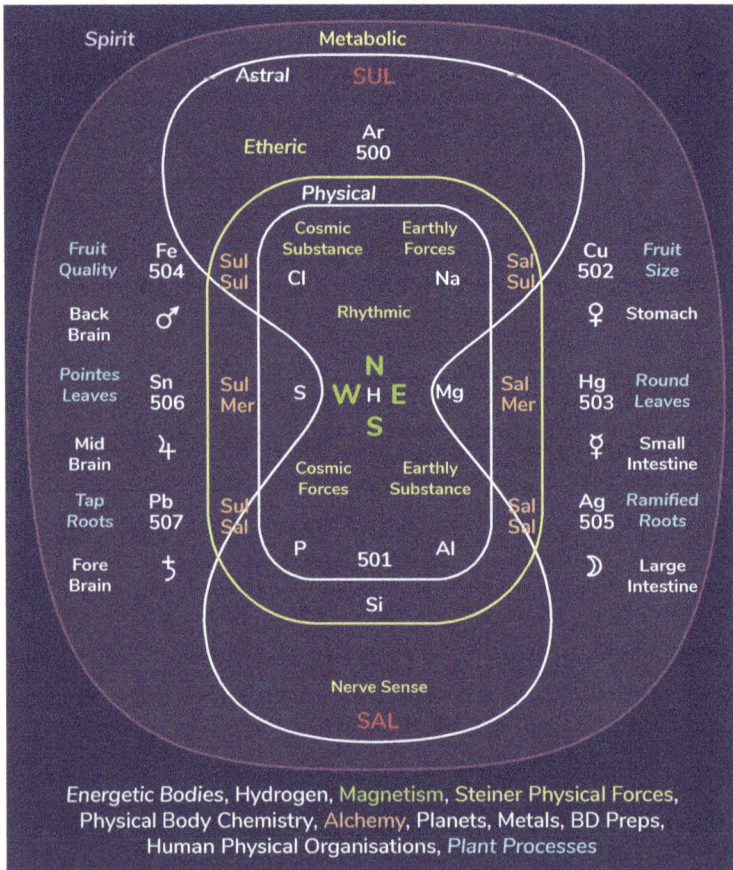

Energetic Bodies, Hydrogen, Magnetism, Steiner Physical Forces, Physical Body Chemistry, Alchemy, Planets, Metals, BD Preps, Human Physical Organisations, Plant Processes

Phosphorus as the Sulf element, and many of his comments regarding it, talk of its function in the nerve sense system, instead of in the metabolic system, as we would expect.

Alchemical Chemistry PT picture

In various later lectures parts of this 6 fold story are told. In lecture 6 (1920) he talks of the planets, then in lecture 7 he talks of how the metals are related to that reference, but the most consistent examples he gives are the relationships of these six activities to those of the head and the digestion. RS emphasised the exact correspondences taking place between these two systems, and that one part of one system can only exist because of the development of the other. In lecture 4 (p. 65) he says *"it is generally forgotten that man exhibits a duality, so that what originates in the lower sphere has always a complementary organ in the upper, and that certain organs of the upper sphere could not evolve without their complementary organs, almost their opposite poles, in the lower. The more the fore-brain approximates to the form which it reaches in mankind, the more evolved does the intestine become in the direction of the process of the depositing of waste material. There is a close correspondence between cerebral and intestinal formation; if the great gut and the caecum did not appear in the course of animal evolution, it would not be possible for men capable of thinking, to arise on a physical basis; for man possesses the brain, the organ of thinking at the expense – I repeat, entirely at the expense of his intestinal organs, and the intestinal organs are the exact reverse side of the brain parts. You are relieved of the need for physical action in order to think; but instead your organism is burdened with the functions of the highly developed larger intestine and bladder. Thus the highest activities of soul and spirit manifested in the physical world through man, so far as they are dependent on a complete brain formation, are also dependent on the equivalent structure of the intestine."* Here we see the polaric relationship of the Moon Saturn relationship in the nerve sense system.

Other stories shine more light on this basic reference. RS makes several comments about the role of phosphorus. Firstly it is within the processes of the intestine that we have the secretions from the gall and the pancreas, which help in the breakdown of fats. Both these organs depend upon a phosphorus chemistry for the production of these excretions. When we come to the head, RS talks of the three sections of the brain, the back brain, middle brain and front brain. He talks of the back brain as the perceptive brain. It is the area that picks up all the 'cosmic imaginations' or the electro magnetic vibrations that are floating past us. We may or may not register these impressions, but generally we are experiencing them as those random visions that come and go across our 'minds eye'. RS comments, that to bring these impressions into clear rational thoughts we have to move them to our forebrain, and this can only be achieved if we have enough available phosphorus (P). It is the P that acts as the incarnating element of the spirit into our nerve sense system. He goes on to describe how P is a 'dissolving agent' and that if this does not work strongly enough then the natural contractive, consolidating and crystallising activity of the head will dominate, and we will experience a sclerosis of the brain. Here we have an image of the relationship of Aluminum and Phosphorus in our head. If the P is not active then the Al will build up and we will experience dementia. One 'cure' for dementia is 'brain gym'. Doing mental exercises like crosswords to keep the brain active. In other words by exercising your rational activity you are increasing the P levels in your brain, so that it can dissolve the Al. So the P is providing the expansive active processes of the Sulf pole into the naturally crystallising Sal pole.

To bring in another one of RS references, we can reflect on his descriptions in 'Man as a Symphony of the Creative Word" where he talks of the metabolic system as imaged by a cow, the rhythmic system is expressed in the cat family, or the Lion, while the head region is expressed in the Birds and notably the Eagle. Thus it can be concluded that people functioning from their back brain – due to high chlorine uptake – have 'cow brain', while strong thinkers express 'Eagle brain'.

As a opposite example to Phosphorus, RS provides an image of salt, sodium chloride, in the 5th lecture (p. 72) *"It is of fundamental significance, that certain individuals in whom the spirit and soul principle is too closely linked with the etheric and physical bodies, have an organic hunger or thirst for salt (NaCl); that means that they tend to reverse the process of depositing salt. They want to cancel the process of earth-formation within their own bodies, and restore salt to an earlier, more primitive, state than that in which the earth has solidified.... And what does this opposition to earth-solidifying forces mean? It means nothing less, in essence, than the liberation of the lower man from the soul and spirit principle, the expulsion of this principle from the lower sphere into the upper in the first instance. Thus in all cases where there is a pronounced appetite for salt, the lower organic sphere is striving somehow for liberation from the too potent activity of the soul and spirit within it, and trying, so to speak, to cause this activity to flow towards the upper organic sphere."*

To add some interpretation to this clause for those not familiar with his terminology, he is saying that the astral and spirit activity in the belly region is too strong, and that the addition of salt **strengthens the metabolic zone** enough, to push the astral and spirit back into the head region, where it belongs. This condition, where the astral and spirit are too strongly incarnated, in this case right into the digestion, will lead to insomnia and also constipation. In insomnia, we can not sleep because the astral and spirit will not separate properly, and thus our front brain thinking processes will not stop. We can also conclude that due to the astral and spirit activity we have too much P activity. Hence we need to push out the astral and spirit, which means we need to strengthen our back brain function, and RS says to do this by increasing the NaCl density.

The constipation reference brings in the story of our digestive system. With the digestion we can identify three parts, We have the stomach, the small intestine and the large intestine. The stomach is the only part of the body where we find Chlorine, in the form of hydrochloric acid. This acid in the stomach has the job of a dissolving the carbohydrate basis of foods. In RS language it is the HCl that needs to 'kill' the etheric forces of the food we are ingesting. This is achieved through an activation of our own inner etheric activity, and so by exercising this 'etheric muscle' we experience an increase in etheric activity. This is a rather strange image I know, but that is what he says. RS says the etheric forces of other beings are too much for us, so we must overcome them, before they are of use to us. It is certainly a simpler image if we just gobbled up other beings etheric forces, but no this is not RS story. So in either case it is the HCl that achieves this dissolving 'etheric' stage in the stomach, and it is the role of salt (NaCl) to control the quantity of HCl we have in our stomach. More alkalinity comes form the sodium and more acid comes from the Chlorine.

The second stage of our digestion occurs in the small intestine. This is where, through Sulphur chemistry, we digest the proteins we eat. Given proteins arise due to the inclusion of the astral body and nitrogen, this is the stage where astral forces are 'digested'. While the P chemistry comes into play with the Gall and pancreas and into the small intestine, it is in the large intestine, that P chemistry is active digesting leftovers and fats, and hence this is the stage of the digestion where the Spirit forces are assimilated. The freedom provided to us by having a large intestine, as a storage organ, is also an image of Spirit activity.

Where there are any disturbances in these processes, we can see that if the stomach activity is too dominant, then the sloppy watery nature of the stomach shows as diarrhea, when the protein digestion is disturbed we have flatulence. The more disturbance the smellier it is, while if the third stage is disturbed by too strong a contractive process dominating, we have constipation.

To add to this picture we can reflect on more information about the 3 fold head system. We have seen how sodium has a relationship to the stomach and a quote from pg 79 in 1920, clarifies its relationship to the back brain, grey matter process. *"It is a complete foolish error to suppose that the substrate substance of thought is mainly given in the grey matter of the brain. This is not so. The grey matter serves principally to conduct nourishment to the brain. It is essentially a colony of the digestive tract, surrounding the brain in order to feed it, whereas the white matter of the brain is of a great importance as substrate substance of thought."* This is only achieved in the presence of Phosphorus.

Where the sodium process becomes too strong, and it dominates its smaller brother Lithium, we see bi-polar disorder spectrum illnesses appear. Interestingly, the treatment is Lithium Nitrate, which we can see on the PT are exact opposites to one another. This process of opposition of elements generally stimulates the activity of both elements.

Where there are disorders in the Magnesium activity, the rhythmic brain is disturbed and we see epilepsy arising, while we saw earlier that front brain disturbances lead to dementia and Alzheimer's, caused by a build up of Aluminum in the brain.

So the solution for insomnia and constipation is to stimulate the opposite pole by using NaCl, while the solution for dreaminess and diarrhea is Aluminum Phosphate. **Alzheimer's could be treated with Magnesium Phosphate**.

These are the conclusions that naturally arise from following RS instructions from the 1920 lectures, when seen from the context of Biodynamics and circular chemistry.

The examples I have given here are from my experience. I have 'proved' these remedies and have found them to act as described.

The Next Step

The previous discussion was focused upon the workings of the elements found upon the Physical ring of the circular periodic table. This is the yellow circle. **By referring to the diagram on page 165, we can suggest that by using the related elements (same arm) to those discussed here, on the other circles, we should be able to identify, and thus influence the other bodies activity, according to the same patterning**. The elements of the green circle will be how we can directly influence the way the etheric body works, in the three physical zones, while those in the blue circle will influence the astrality, and those in the purple circle will influence the spirits activity in the three zones.

I have done some experimentation into this association, and find it to be a worthwhile avenue of exploration. My experiences conform with what I would expect to occur. No doubt the future edition of this book will contain more experiences. For now this 'indication' is enough.

The correspondences offered here, through both the World Physical and Internal Physical orientations, should provide a complete reference for the use of all the chemical elements, within the context of Dr Steiner's worldview. As much as I would love this to be a completion of the task of extending RS indications on chemistry, no doubt there will be some more 'wait there is more' moments ahead.

While the World Physical orientation provides the references for identifying the many levels of a elements energetic activity, this 'Alchemical' extension of that work, brings these activities right into the bodily processes. Thus providing an avenue for more very practical applications.

The Seasonal Complex

In addition to Dr Steiner's medical lectures we can bring in three lectures from the agriculture lexicon, that links Alchemical Chemistry to the seasons. The result is what I have called the **Seasonal Complex** (p. 174) I have presented a more detailed story of this as part of my 'RS Plant Growth Story'. The three lectures I consider pivotal to understanding RS agricultural worldview are

Cosmic Workings in Earth and Man, Lecture 5, given on October 31 1923
Man as Symphony of the Creative Word, Lecture 7, given on the 7th November 1923
Agriculture Course, Lecture 2, given on 10 Jun 1924.

All these lectures provide an image of the forces active throughout the annual growth cycle, however interestingly they each use a different reference language to each other. Hence they need to be bought together, and seen as one story, as best we can.

The first lecture is from "Cosmic Workings in Earth and Man", lecture five given on the 31st of October 1923. This lecture talks of the SAP story regarding plant growth, and it has three primary players the 'Wood Sap' coming from below, the 'Life Sap' coming through the light activities, which make the leaf and foliage. When the Life Sap is strong enough, it draws in the incoming 'Cambrium' force, sourced from the Stars. This stream is carried upon the outer planets 2 frequencies, and through the cosmic warmth processes, along the cambium pathways of the plant, and in towards the Centre of the Earth, where the World Spirit of the plant kingdom resides. During the winter, a maturing of the Wood Sap in the soil, combines some of this 'earthed cambium star forces', with mineral earth forces, and together they push the Wood Sap upwards, in Springtime.

Alchemical Chemistry PT picture

The second lecture from "Man as a Symphony of the Creative Word", lecture seven given on the 2nd of November 1923. This is some three days after the previous lecture, and here Rudolf Steiner talks of how the Ethers work through the seasons, and also how the Elemental Beings work as transformative agents in the development and synthesis of these Ether processes. He outlines the flow of activities within the Earths Ether body. Using the Earths surface as the diaphragm membrane between the Above and the Below, we are told how the different Ethers move above the surface of the Earth, and thus have a direct effect upon nature, in one hemisphere. Before they recede again back within the Earth, and become active there and in the other hemisphere. From my experience, I suggest this movement can be seen in nature. RS also tells of how the Elementals play into facilitating the interaction of the primary forces, be they Ethers, Saps or Cosmic and Earthly Forces and Substances.

The third lecture in my series is the second lecture from the Agriculture course, given on the 10th of June 1924. This is some 8 months, later after these earlier two lectures, however it presents the third view of energetic plant growth. In this lecture we are presented with a view of the plant sitting within what we call 'the threefold agricultural individuality'. In the beginning of lecture Dr Steiner clarifies for us, that the plant in the Earth is like a human turned on their head. So the nervous system is in the Earth, and the metabolic system is above the plant. Above the leaves to the top of the plant, and even into the atmosphere around the plant, is considered the metabolic region of the plant. The leaf region is the Rhythmic system, equating to the human chest region. This threefold image is an archetypal organizational structure that we find throughout the physical bodies of all manifest beings.

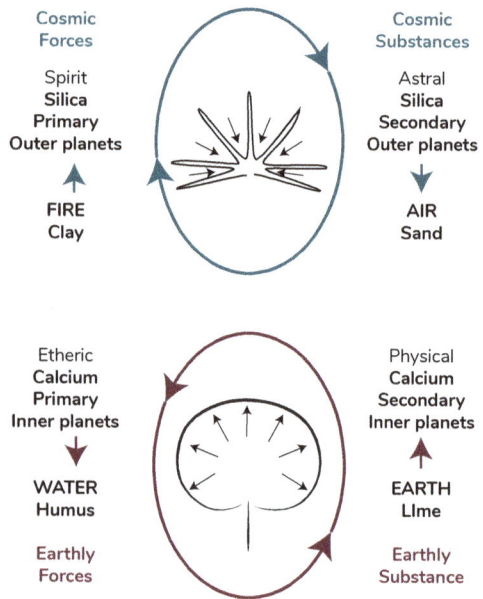

This story only takes about two pages to tell, and so the rest of lecture talks about the Physical Formative Forces (PFF). Most readers of the course remember the line 'the ABC of everything is what is Cosmic and what is Earthly'. This is a pivotal story, however many students only get so far is seeing Cosmic and Earthly as a simple polarity, between what is above and what is below, and carried by either by Calcium and Silica. In the second part of this lecture, we have an intricate story of the Calcium and Silica processes working as Forces as well as Substances. This gives 4 processes to consider. Cosmic Forces and Cosmic Substances, along with Earthly Forces and Earthly Substances. Each pair form a circular process. One part of the process comes towards the Earth and the second part is moving away from the Earth. This occurs in both groups. The real gift of the second lecture, is RS telling us the physical carriers of these four activities, which are available for us to use. He talks of the Cosmic Forces being strengthened by the Clay substance, the Cosmic Substance is strengthened through the physical presence of Sand. The Earthly Substance is strengthened through what he calls Lime, which is also to be taken as cations in general. Then the Earthly Forces are enhanced through the activity of humus, which is more specifically carbon and all the processes associated with carbon chemistry in agriculture.

This story is told within a seasonal context, so can be related back to the previous two lectures. But within this lecture we are also provided with an image of how these physical formative forces play into the processes of plant growth, as controlling levers. In the earlier lectures, we weren't given any particular preparations that we could use to control those processes. However with this lecture we are given the four basic elements for 'levering' the seasonal flow of the energy, and so influencing everything talked of in these three lectures.

The rest of the agriculture lectures talk more about the physical formative forces level of manifestation, along with stories of the primary energetic activities, sourced from the World Spheres. Interestingly, there's only one significant reference in the agriculture course to the Ethers. The great majority of the course talks of these two other levels of activity. This appears to be why the people who use the Ethers as their central reference for understanding biodynamics, have great difficulty understanding the Agriculture course. Their story is told in the 'Man as Symphony' lecture series. I see people have studied and worked with the first two lectures, however I do not see a lot of literature or work done on the Physical Formative Forces story. It is a different story to those, however extremely compatible with those other stories, and now we have some physical levers, by which we can control all three stories. The biodynamic preparations provide more levers working through the primary energetic activities, which do work into and through the lower levels of the Ethers and PFF.

My picture of the 'seasonal complex' is where I integrate these three stories into one dynamic.

The Seasonal Complex

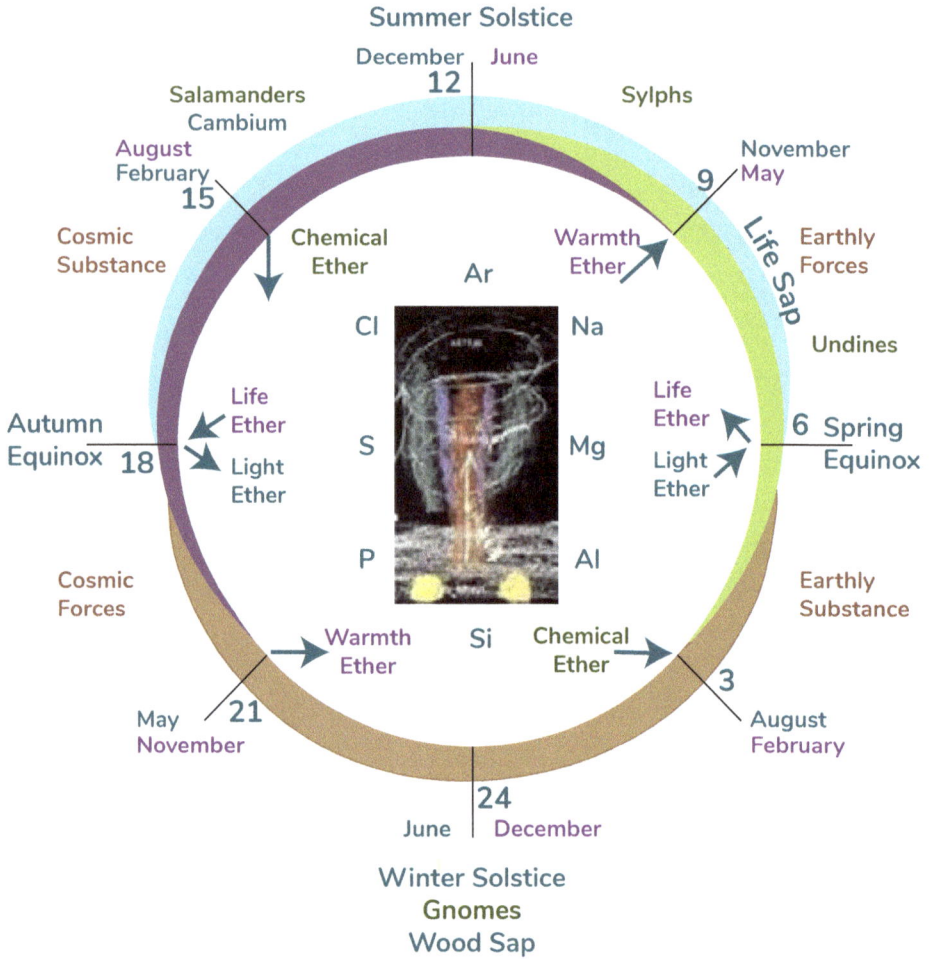

Summer Solstice

December | June
12

Salamanders
Cambium

Sylphs

August
February
15

November
May
9

Cosmic
Substance

Chemical
Ether

Warmth
Ether

Earthly
Forces

Life Sap

Ar

Cl   Na

Undines

Life
Ether

Life
Ether

Autumn
Equinox **18**

S   Mg

Light
Ether

Light
Ether

6 Spring
Equinox

Cosmic
Forces

P   Al

Earthly
Substance

Warmth
Ether

Si

Chemical
Ether

3

May
November

21

August
February

24

June | December

Winter Solstice
Gnomes
Wood Sap

Walter Russell Sub Atomics

A fellow journeyer along the way, is Walter Russell who was an American genius of the 20th century. He had many talents and one of them was Chemistry. While he achieved marvels, such as changing one element into another, he also predicted the place of some elements. He proposed that there are four rings of elements, below Helium. Sub Atomic elements. Charles Walters puts forward an association of them, to the recognised sub atomic particles.

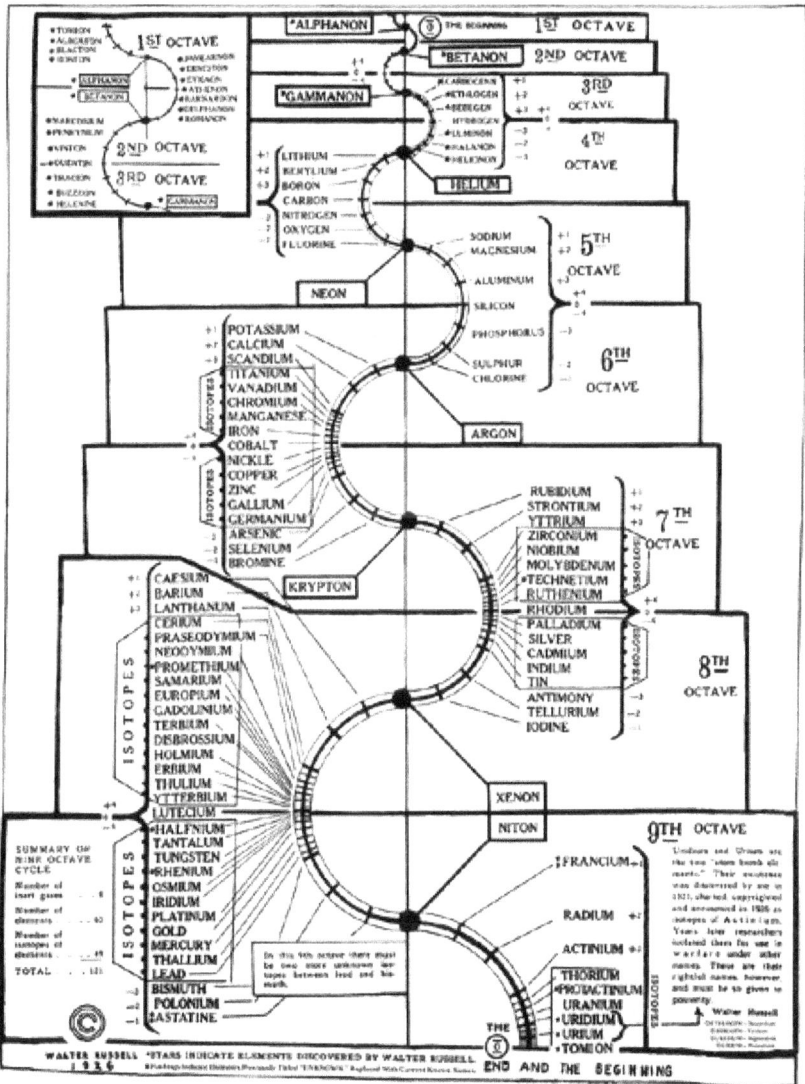

The Russell Periodic Chart of the Elements, No. 1

Similarly to Jan Scholten, WR's chart is drawn as a flatten spiral vortex form rather than circular. At the top of his chart, on the left corner are the sub atomic elements.

Here is my suggestion of how to approach these elements and what their possible functions and uses would be.

Phase 1 has the red circle in the middle of my normal Periodic picture, enlarged to allow for WR Sub Atomics.

Walter Russel Sub Atomic Particles
Circular North

Phase 2 is developed when we apply the law of polarity, which suggests that once we enter, into the sub atomic level, which may well be like going through a black hole, we find a mirror image of the order found in the exterior world.

176

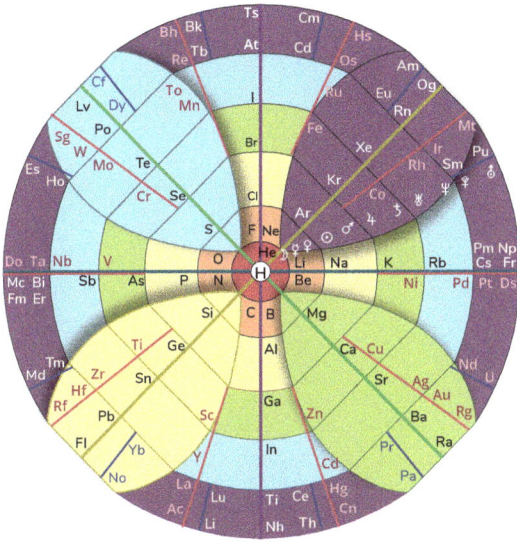

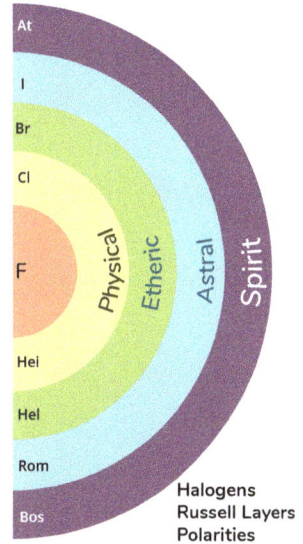

Halogens
Russell Layers
Polarities

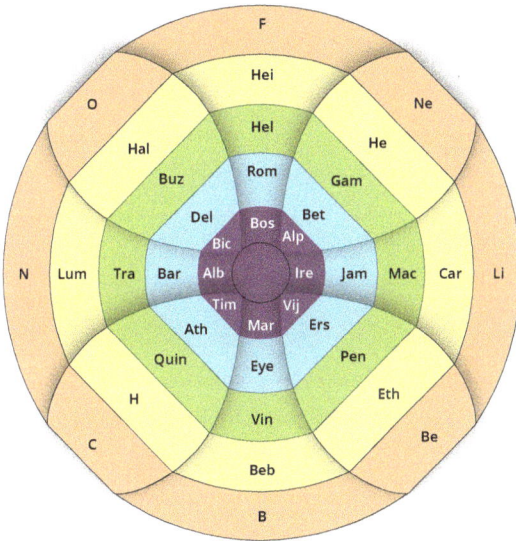

Walter Russell Sub Atomic Particles
Circular North
Layers Polarity

There are 9 Layers
Which may Polarise?

4 Shells with a Centre
The Earthly Substances –
The Sphere of Polarity,
from which Life Emerges,
gets to play this Role.

Walter Russell
Sub Atomics
Circular
Layers Polarised
Circular Periodic

Phase 3, Alchemical Chemistry says we can spin the last picture 45 degrees to match the Alchemical Chemistry pictures, and see if the same rules apply.

This all suggests that if we are to **use an exterior element, we can also use its sub atomic companion, together**.

All the same issues about accessing and collecting chemistry.

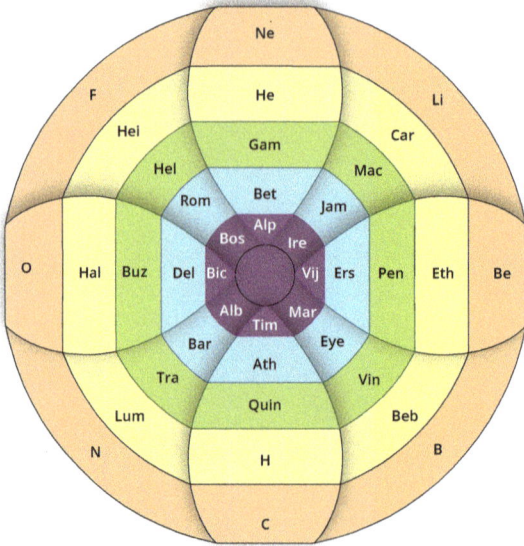

Walter Russell Sub Atomic Particles
Circular North
Layers Polarity

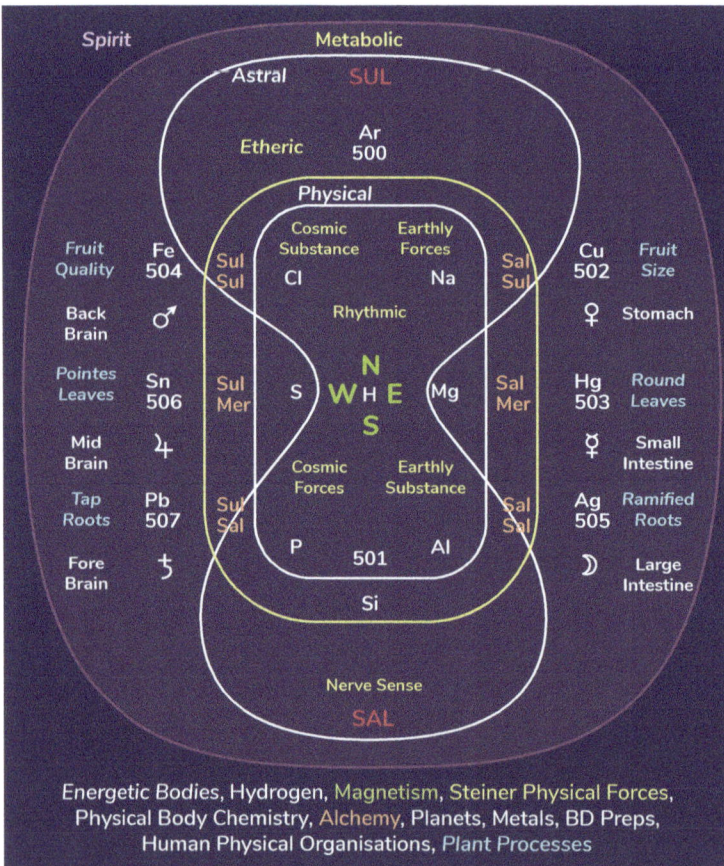

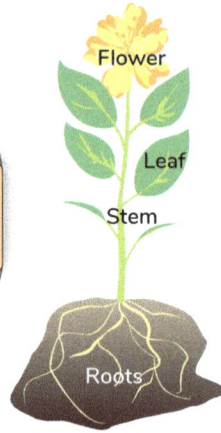

Energetic Bodies, Hydrogen, Magnetism, Steiner Physical Forces,
Physical Body Chemistry, Alchemy, Planets, Metals, BD Preps,
Human Physical Organisations, Plant Processes

Practical Considerations

One of the largest problems of explorations into chemistry, is that most of the chemical elements are either toxic to handle, explosive when exposed to air, or radioactive. Thus one needs to be relatively highly trained chemist to not kill yourself.

The second 'problem' of chemistry is the availability of these toxic and dangerous substances. Basically the average person can not access many of the elements, and then of course one can only access those that naturally appear in nature. So there are various combinations one might like that are not available at all.

These are certainly large enough barriers, for the insights available from this book, to remain intellectual potentials, but practical impossibilities. However there is an answer to these problems, and one that may allow the insights here, to be available to all of humanity, free of charge. It is this very ability of this system, that 'forces' me to give this whole methodology freely to humanity. As now more than ever before, the people of the world are in need of a non-corporatised answer. Yes there is some effort needed to bring all the information together, but time and co-operative efforts by interested parties, should provide a simple materia medica, normal people can use for simple solutions to their issues. Homeopathy has a lot of information about the chemical elements, already.

The answer to the above questions lie in a few simple truths. A cornerstone 'truth' of all of my efforts is that Creation = Movement + Time. From this movement over a very long period of time, creation has unfolded according to basic principles that can be identified at all levels of creation. Many people have shown that the mathematical image of this order is the Golden Mean, while RS, and others have reaffirmed the axiom of 'As Above, So Below.' My entire body of work is dedicated to showing the reflected order that exists between what is above us, with how life process organise themselves. From this, it can be seen that there is archetypal order, from the beginning of the story as expressed in 'Biodynamics Decoded' through to the end as expressed on page 48. It is important to observe, that my entire story follows one stage after the other, from simple observations of what is there, as scientific fact in front of us. I am seeing what is just there.

A cornerstone recognition for me was when I began to travel, and noticed every culture on Earth has revered the circle and octagon within it. Why is this? Because this form is the representation of the macrocosmic truth that we live within a electro magnetic sphere, where all matter organises according to the North South polarity of the strongest field. This sets the external reality of life. We also identify that internalised life, is the below, and a reflection of this reality. Thus the second cross, which makes up the octagon, and related to the Earth based equinoxes and solstices, provides an archetypal image of the creation we are. **It seems all cultures have recognised that the simple act of drawing a circle, within the electro magnetic soup of our life sphere, brings about a 'being' that can resonate with the archetypal form of the energetic sphere that is the Galaxy, Solar System and Earth**. The circle becomes a energy organisation device, and the pattern of organisation is the order I have described throughout my writings. All circles and octagons will create this order, regardless of their size or the material from which they are made. The older cultures know this. There is a great story of the large stupa in the middle of Katmandu. There is an army of 300 people protecting this stupa, as they believe and no doubt experience, that the stupa generates a force of cosmic harmony that radiates across their lands, and creates stability for the development of their culture. Their ongoing peaceful existence depends upon the existence of the stupa. Islamic culture identify this form also as a representation of cosmic harmony, and all their major buildings are octagonal.

One of the best examples of the structural use of this form is St Peter's Basilica and Square in Rome. St Paul's in London is another great example. The christian's made a slight change to the basic form of the Greek cross, by extending the west apse of their building, but the basic Greek cross is still formed by the equal lengths of the other three apse, which then resonate the boundary of the fourth side, often a third or half way down the western apse. The clue of the Christian cathedrals is all of the Greek cross imagery in the tiles and paneling throughout the buildings. The priest garments also display it. There are several examples of octagonal chapter houses in the later Templar cathedrals.

Other cultures such as the ancient Britain's made stone circles, while the American Indians call their earth drawn circles 'medicine wheels'. They have very clear understandings of the effect of the various parts of the circle. The detail within the Tibetan mandalas formed on this ordering, suggest they too would have detailed understandings of the internal structure, as do the Chinese with their I Ching and Fung Shui.

So while the knowledge of the power of the circle and octagon is lost to our western culture, it has not been lost to several others. My efforts, firstly by just following the 'As Above, So Below' principle of astronomical order, and then by 'pulling at the frayed remains' of what has been left lying about my cultures past, has recreated a precise map of what activity can be found at any spot in the circle. Simply by standing in any spot of an Earth circle, one can experience and be effected by the specific energy gathered there. We only need to see the experiments of sand on a vibrating plate, to observe the process. The form of the sand arises from the vibration. By changing the vibration the form changes. Just as the form arises from this vibration, so when a circle or octagon is drawn, the Earths magnetic field drives the universal organisation to occur.

This occurs with every circle, or octagon of whatever size. So once you notice these lying about your environment, you can see an opportunity to find whatever remedy you might need, literally at your finger tips.

The picture on the cover of this book is an energetic organisational device, that can be used simply by putting your finger on any energy you wish to attract, to yourself. Be sure and align the north pole of the picture towards the north in your environment.

The advantage of this method is that at no time do you come in contact with the physical chemical substance, and therefore all the problems of dealing with their atomic weight do not exist. Nothing will blow up, nothing is radioactive, and I would like to say nothing is poisonous, however it is possible to get too much of any particular vibration, and this could cause an imbalance, that could cause some illness to occur. So 'tread carefully', it is a art form that needs some practice.

Collection Method

While standing on a spot or even using a finger on a spot, can achieve energetic movements, it is also desirable to be able to make up remedies that can be taken along with you and used as needed, and given to others.

We have the wonderful natural qualities of water and Silica at our service. Dr Omoto of Japan, and others, have shown that water is a programmable crystal. Simply by bringing water into a persons field, the water will take on whatever energetic quality the person is concentrating upon, immediately.

Glass bottles are made of Silica and we know that silica has the ability to transfer electro magnetic energy. **Therefore water inside a bottle and placed in the Sunlight, while placed upon a specific spot within the circle, will accumulate and hold the energy of that spot.**

Thus you now have the ability to collect whatever energy you want. This remedy type is essentially very similar in quality to a Bach Flower remedy, which is essentially made in the same manner, except flower parts are placed in water in the Sunlight.

The individual remedies can then be combined or they can be potentised to further specialise their effect. However, now we can use all the elements of the periodic table, I wonder if potentising is necessary.

Collecting Sub Atomics

One of the problems with WR sub atomic theory was that they are energy nodes, rather than real substances. Within the circle though we have them placed in the first circle in the centre. So if you have a big enough circle then the centre isbig enough to identify 4 rings within that zone and by placing bottles on those spots you can collect those energies. I used a 4 meter radius circle, which allowed each of the 6 rings to be 666 cm. Thus the four rings inside the inner most circle were 166 cms each. This is big enough for a good size bottle if you wish. I place the bottle in the centre of the zone.

In Steiners' methodology, the substance is selected for its activity upon the energetic bodies, and then the potency is used to direct it to the physical system we what. We need to remember he and his followers have only described the function of some 20 chemical elements, and while they did publicly present the framework I have shown, it appears they may well have had my Glenological Rosetta Stone images.

How to use this information

In previous chapters I have outline several approaches to seeing the organisation of Chemistry, and while I hope you will be able to see the practical applications that arise from these differing views, I suspect I am doing us all a favour by outlining how I use this information.

The basis of this whole system is that manifestation occurs as an interaction of the four energetic bodies that find their source in the four great astronomical spheres, as outline at the beginning of this book. The best metaphor I have found for how the bodies act is, what is involved in building a house. The Spirit is the architect, who has the plan. The Astral is the master builder who has the energy to make the plan happen. The Etheric are the workers who follow the direction of the architect, but ultimately follow the orders of the master builder. Then the Physical body is the wood and gibboard etc used in the building.

The Spirit brings order. Through a inward moving motion pushing to the centre. The Astral brings energy and movement. Also having a inward motion, however there is also the tendency, when not controlled by the Spirit, to spin things outwards, giving a sense of being 'out of control'. The Etheric brings the fuel to do the job, and has an expanding enlarging motion, which can push off the Astral as well as the Spirit, if the Etheric is working too strongly, providing an experience of 'mummy brain'. Each needs each other. Without the Spirit there is no direction or context to work within, and without the busy Astral, the Etheric becomes lazy and stops working, to become 'stagnant'. The Physical also has a contracting influence bringing everything into manifestation. The Astral boss has to follow the Architects direction and drive the workers, to form the substance into the appropriate place. It is this simple.

The Energetic Circle

The first step in this study was the identification of the energetic organisation of the Circle. This developed out of an artistic process of looking at astronomy, and how it is organised, giving the Rings image. Then learning of RS's suggestion of what happens to the primary energies when they move and form into biological entities, we have the double cross image. In the search for a way of portraying RS's ideas of how the energies interact with each other, I placed the astronomical 'Rings' reality over the biological 'Crosses' reality, which gave the required image of every activity interacting with each other. This then allowed for the Agriculture Course language to be placed in their appropriate place. Providing the images on page 48.

I accepted this as a artistic process for a around 10 years, until I became aware of the reality that humanity has known for over 10,000 years. that circles are universal energy organisation devices. Everything in creation is an energetic sphere and circles are just one slice of a sphere, so every circle therefore MUST resonate with everything else in creation. Using the Earth magnetic field as the organising 'power', we should therefore be able to find THE universal organisation within EVERY circle. It took another 10 years of living with circles and being sensitive to the changes in energy within the circle, to conclude that the organisation I have found through the artistic process just mentioned, was indeed THE universal organisation. It is important to note, most northern hemisphere circle systems are based upon the Sun rather than the Earth magnetic field. I choose to work with real magnetism which effects the real chemical elements.

A practical use of this finding is that we can accumulate any energy interaction we might like by standing in the appropriate place within any circle, once we have determined where Magnetic North is. It is also possible to place a bottle of water in the desired area for more than 10 minutes and collect a remedy you can take with you for future use. For the science of using the energetic activities I can but refer you to the Steiner medical lectures and the subsequent Steiner doctors medical literature. One lecture series I found very useful was one of RS last. Pastoral Medicine (GA 318 Sept 1924). This provides a fairly clear picture of what occurs as the energetic bodies change their relationship.

Biodynamic Chemistry (pp. 63 – 144)

The main developments that arise from this study is the identification of the energetic activities of all the chemical elements. This is best illustrated in the diagrams on pages 152 –153.

Observation and experience over the last 20 years, by myself and several others, most notably Hugh Lovel, have shown that the associations made through this study are correct. The 'predictions' of effect made in these diagrams have proven themselves many times.

The Diagnosis Process

I will relate some stories of treatment of dogs, as this overcomes questions of human subjective bias. The reference for the energetic activity of the animal kingdom is Dr. E. Kolisko and his 'Twelve Groups of Animals'. The picture he gives is that we start with single celled animals, as an image of Etheric dominant influences. Once we see the indentation of the gastula we are seeing the incarnation of the Astrality. As we move through the various phylum we see an increasing incarnation of Astrality, until we see the emergence of blood, which is an indication of a Spirit activity being present.

The following experiences come from Linde Walker, an experienced dog groomer and carer. She had a 15 year old dog with a mite infestation in its paws.

Mites are an insect, indicating an astral problem. They suck blood, which indicates some aspect of the Spirit is involved. This problem suggests a Spirit Astral imbalance. Given the dogs age, we can expect the Spirit to be separating from its bodily functions, just as we see in aged humans. As the Spirit leaves the Physical, Etheric and Astral complex of an animal, its directive powers upon these bodies weakens. Astral forces start to act on their own, which in a Human shows psychological and emotional disruptions, along with susceptibility to bacteria and insect infections. Various bodily shaking can also appear. The Spirit orders and controls, pressing inwards.

So we want to strengthen the Spirit and weaken or calm the Astral. This is a living being so we are concerned with the Internal activities. So the Internal

Astral area – blue petal – is indicated. To bring the Spirit activity into this we can focus on the Cosmic Spirit ring, which is the outer purple ring. There are 8 possible elements identified by doing this. Holmium Ho however was my choice. Mites being a tiny animal, it suggests we go to the start of the animal area.

Another study I did is to be found in the 'Circle of Everything' picture. This showed that Koliskos 12 groups of Animals can be found placed around this same 90 degree, Astral quadrant. The first 6 phylum are placed along the World Astral horizontal divisions, while the other 6 are equally spaced from the horizontal up to the vertical axis. This places the Insects, the 7th Phylum about 15 degrees up from the horizontal axis. So the closest element to this spot is Holmium, a Lanthanide. The whole Lanthanide group is associated with helping the internal Spirit work, so another tick in this box.

We can also note that Holmium is ruled by the planet 'excarnating manifest Mars'. Excarnating processes appear to be associated with 'manifest' things like insects etc, while Incarnating process appear to be more active in bodily processes, as we will see with Samarium – 'Incarnating Force Mars'.

This diagnosis of Holmium will also relate to similar problems like ticks, mosquitos, sandflies and fleas etc. This also suggests that the corresponding Mars 2 element on the trace and majors groups could also be used for these problems.

Here is Lindes account – "I used the Lanthanide Ho (Holmium) on a very old 15 + Shihitzu who contracted Mites which were causing them to eat into his body and in particular the ends of his tail. Anyone who's dog has contracted these Mites will know they are very very hard to get rid of. I am care-giving this lovely old Soul and so decided to try Lanthanide Ho. On the emotional side it helps to give clarity and direction and helps with things like positive enthusiasm, attending necessary details, expanding, willingness to take some risks and benefits of long term goals. This little Soul can at times (especially at night when we turn out the lights) goes into ANXIETY and stresses and hypo ventilates and paces around. However, just after the SECOND day of taking Ho Lant he is relaxed and... no more anxiety. I plonk him on my bed and he settles straight away off to sleep. The mites which had gotten right into the end of his tail is now clean flesh, he can feel his tail more, he has new hair follicles growing back and all that horrible black stick mite exponent has completely disappeared. I am so excited. This is a very natural method of dealing with mites, fleas... and... helps with their own emotions because Ho brings them back into their own spirit... this makes a happy chilled out puppy at 15+.

I have just finished an experiment on another two little 'Oodle' breeddogs. One has chronically suffered from yeast and bacteria. So these patches come up on their skin and in the yeast one it just smells so pungent and moves to their inner ears. They other little Oodle (as in Poodle mix) had more of a bacteria than yeast patch on his side. I prescribed 2 drops per day of Ho for a week and....to apply topically as well so as to attack this spreading bacteria. The outcome, was astounding. Any and all bacteria and yeast is now completely gone except Lottie has (chronic in the past) and still slightly got yeast in her ears. I have asked if they can put 1 drop per day in her ear to hopefully kill this off. this is exciting news to me. If this works on my little Furbbies, then....what about farming animals, especially organic farming animals rather than using chemical, toxic methods of healing. I will tell you the side affects of my little experience

The side affects of both Lottie (female) and Harry (Male) is, they have a new Retriever puppy living on their site. Harry has 'NO' tolerance for her cute boisterous puppy play and is (was) aggressively putting this poor excited large puppy in it's place which was causing her to submit too much. Daily when put out to play Harry would bite her ears till she squealed, bite her legs. Lottie would rush her as this little female puppy was coming between Lottie and her husband Harry. Lottie is usually the sweetest gently little soul. However, due to both Lottie and Harry being on Ho for mites/bacteria/yeast and being that Ho is Mars/Scorpio and help Humans (plus dogs it would appear) to 'get on doing things with positive enthusiasm, making things work, clear head, things are in order,' that their Mother tells me, with the most astonished reality that, whilst Harry normally aggressive that he stopped being aggressive, is now welcoming and playful with the puppy. Lottie is now beginning to cozy up to the Puppy when they rest and sleep and cuddle each other. Wow! Fantastic results.

So far the experiment of combating fleas has worked. Not one flea on both dogs.Glenand I are now wondering how long will it take before fleas appear again before having to give them (2 drops 2 x per day for 2 days or put the drops in their drinking water). 4 weeks have gone by and all good so far."

A different example, this time with Neodydium (Nd). This is a Lanthanide, so the outer purple circle, but in the Internal Etheric petal. The Etheric has the effect of pushing off a too strongly acting Astral, so has a calming effect on psychological issues usually.

"Fantastic news. The Pitt Bull X that was soooo unruly and aggressive and preditory. The new owner was. "Can't control this dog. Need her gone." So we re homed it to another new owner. I sent Nd Lantz to her. I was just informed Mala (the dog) is doing "spectacularly. I am so ecstatic and pleased for all otherwise it was the needle and goodnight Nurse. (Nd is the venus 1 element – which emphasises the social skills needed to bring your goals into fruition. So it makes you socially safe and willing to 'be friendly'.)"

Samarium

The difference between 'incarnating and excarnating' planets are best seen as part of a circular process. Energy comes at us from the stars, it moves through the planetary spheres, then through the Atmospheric shell of the Earth before being received by the Earth. The Earth then reflects all this combined energy back outwards as life forms. So we can talk of the incarnating processes as Forces, while the excarnating processes show as Manifestation.

Samarium is Incarnating Mars or Mars 1, and my experience with it shows it has many uses. Mars is the planet of the Astrality. Saturn is the planet of the Spirit while Jupiter helps the Spirit and the Astral work together. So Mars in the Internal Spirit region suggests a place of how the Spirit and Astral work together, to organise subsequent manifestations.

However it is working on biological functions, rather than exterior pests. Mars is the God of war and governs the adolescent male, so hot headed impulsive actions. I first used it after I had a heart attack and was experiencing hot flushes. This is because the Internal Spirit is not properly incarnated due to the heart attack process. The hot flushes stopped. I then gave it to some menopausal women, who found it helped them with their hot flushes. Next, as a 'leftover' from working a lot with stinging nettle – which is a astral stimulating herb – in the late 1970s, I come out in a rash on my arms, each November. If I scratch it it gets worse. Over the years I have tried many things, and cold water is as good as anything. Once I started applying Sm onto it topically, the itching stopped. I have since taken it as internal drops and this has worked even better. Recently this treatment has meant this annual event is no longer a problem.

On another occasion we were camping. and I was collecting some Sm from a nearby circle, when two people got stung by wasps. My Astral Cooler product, which contains stinging nettle has proven very useful against this in the past, so I thought Sm was worth a try. Within a couple of minutes all effects of the wasp sting had gone. It was also a successful pain relief, when drops are rubbed over the painful area.

The message I am trying to convey is that once you have identified the energetic activity you wish to work with, you can identify the chemical elements that work with those processes. There is often two or more places where you can find what you want.

In the Holmium example – Cosmic Spirit, Internal Astral – where we are wanting to bring the Spirit and Astral into harmony, we can also look to the Internal Spirit petal and the Cosmic Astral ring which indicates both Xenon and Rhodium. Either of these can be combined with the Holmium if you wish. Chemistry is rarely just one single element it always has a cation and anion pair, so I like to have a partner, however this is not always necessary.

Steiner's Organisms

A good example of pairing elements is when treating Steiner's organisms. I just mentioned the hot flushes I experienced after a heart attack. RS would say that my warmth organism was not under control.

An 'organism' is created when one or other of the bodies works into the physical body over a period of time. When the Spirit works into the physical body. It creates a sheath within the physical body, that controls how we manifest warmth or not through out our body. When the Astral body works into the physical body it creates a Air organism, which shows up in our breathing processes. So Asthma occurs when the Astral body upsets the Air organism in some way. When the Etheric works into the Physical body we have a water organism, which controls the flow of fluids through and out of our body. As we age the Etheric weakens it link to the Physical and we have incontinence. The solution to all these problems is to bring the various bodies back into deeper penetration of the physical body again.

With the warmth organism we want to bring the Spirit into the Physical, so we can look to the Cosmic Spirit ring in the Internal Physical which is Lead (and it associated elements), while we can also have the Internal Spirit with the Cosmic Physical ring which is Argon. So PbAr.

When the Internal Spirit excarnates, the warmth organism continues to function within the physical body however it has lost its controller.

For the air organism we want to bring the Astral into deeper penetration of the Physical. So the Internal Astral petal with the Cosmic Physical ring is Sulphur, while the Internal Physical and the Cosmic Astral ring is Tin Sn. So SnS can be used for breathing issues.

For the water organism we have the Internal Etheric petal with the Cosmic Physical ring, which is Magnesium Mg, and the Internal Physical and the Cosmic Etheric ring which is Germanium Ge. So to control your water works you can use MgGe.

The myriad of health issues we are confronted with can all be addressed using this system once you have the energetic activity interaction identified.

Potentising

In this method we have identified the energetic activity we wish to use however we have not directed it at a physical system. This is achieved through potentising the 'substance'. RS's suggestion for this is the potencies from 1-10 work on the metabolic system, 11-20 work on the Rhythmic system, 21-30 work on the nerve sense system.

The next question is which exact potency within these ranges. There are two ways to do this. One is to follow Dr L Kolisko and grow wheat plants to the third leaf stage, measure the length of the leaves and the root, through all the 30 potencies. Plotting them on a graph, and then make assessment as to which potency provides the strongest influence in the group. I did this for my BD preps.

The second method is dowsing. This is the only time I use dowsing. I do not use dowsing for diagnosis, as one of RS's gifts is the ability to consciously diagnosis energetic manifestations, as shown above. This is a real 'muscle' and if you do not exercise it you will not build it, which means you are not becoming a conscious creative 'angel'.

Firstly I ask the universe, which potency between 1-10 is best for this substance to do whatever job in the metabolic system. Usually I get number. Then I will move my finger over a drawing of a scale of 1-10 and ask the same question, to check. I experience a little bump at the right number. More often than not they are the same. I appreciate both of these are of 'dubious scientific repute', however all the plant trial work is quite a challenge for most lay people. I called a halt to it once I started working with chemistry and its over 100 elements. Just far too many bottles.

To potentise something you take 1 part of the substance and dilute it with 9 parts of as pure water as possible. Distilled water and then rain water are preferable. Only fill the bottle around 75% of its capacity. Then rhythmically shake the bottle for 2.5 minutes. This is D1. Then take this amount and dilute to 9 parts, and do it again. This is D2. You can take only some of D1 and dilute that amount by 9 parts and then shake that. Carry on this process till you reach the potency you wish to use.

For the chemistry elements I now use a simple radionic potentising machine, as there are too many bottles involved otherwise, and it appears to do the job.

Incarnating and Excarnating Planetary Influences

In the previous passage by Linde she referred to Mars 1 and Mars 2. This is referring to the process around the circle. I covered the details of this in the Journeys through the Lanthanides. So you can identify the stage of the cycle you are interested in and then you have the option of which Ring of the circle you wish to use, so you can focus upon an energetic body activity. The diagram on page 138 shows the planetary rulerships.

World and Internal Groups

Seems to me that the Internal and World elements work in a similar way to the Force and Substance sides of the circle. Force or the Incarnating planets side work on processes, while the Excarnating Manifest side of the circle works more on manifest problems like insects. Similarly the Internal arms work on Manifest and inner cellular biological functions, while the World Activities work on more general extra cellular processes, as well as the outer environment such as soil and weather.

I am not a good enough biologist to be able to define the details of this difference clearly.

Walter Russell Sub Atomics

My study of these elements suggests that the Sub Atomics form a mirror image of the manifest elements, and that the 'opposite' elements work together or individually to create the same effect. So if you wish to strengthen the Internal Etheric body into the Internal Physical Body, an action that supports good health, then you can use Magnesium (Mg) and Ethlogen (Eth) This is as if you played the Note of E in the 4th and the 6th octave, at the same time.

The Seasonal Cycle Treatments

In 2019 I ran a trial to see if I could use the seasonal cycle as a an indication for remedies appropriate for the time of year. It was of mixed outcome. One of the reasons I wrote it up and made it available was to show the thought processes I used to come to my suggestions.

It is now 2022 and I have continued my investigation on this 'idea', and have come to a new understanding of how we might work with this. Mostly due to efforts I have made to control the 'rash' that surfaces each year in November on my arms, I had found that Samarium applied topically would reduce the itching. Now when I have extended the mineral references within the Seasonal Complex picture I find that in November when the Warmth Ether is becoming more expressed in Nature one of the indicated elements for this time is Samarium. This year I have taken it internally, once every few weeks and I have only a very slight rash. This suggests this reference may well have great value for helping with seasonal changes.

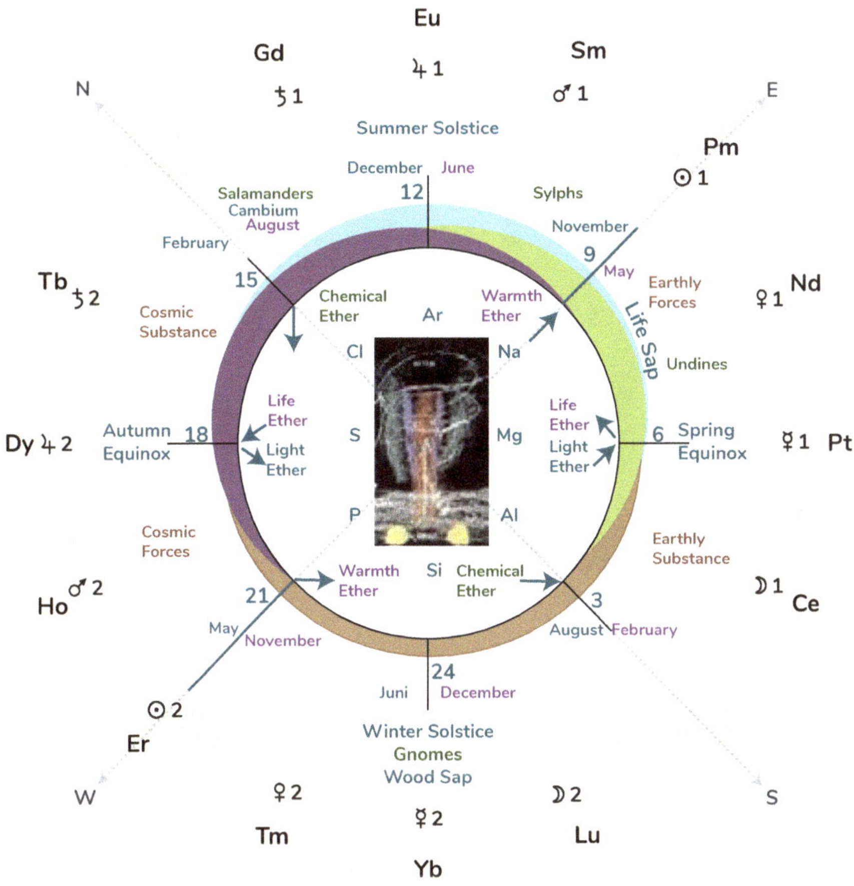

Eu
Gd
Sm
♄1
♃1
♂1
N
E
Summer Solstice
Pm
☉1
December | June
12
Salamanders
Cambium
August
Sylphs
November
February
9 May
Earthly Forces
Nd ♀1
Tb ♄2
15
Cosmic Substance
Chemical Ether · Ar · Warmth Ether
Cl · Na
Undines
Life Ether · S · Life Ether · Mg
Dy ♃2
Autumn Equinox 18
Light Ether · Light Ether
6 Spring Equinox
☿1 Pt
P · Al
Cosmic Forces
Warmth Ether · Si · Chemical Ether
Earthly Substance
Ho ♂2
21
3
☽1 Ce
May November
August February
24
Juni | December
☉2
Er
Winter Solstice
Gnomes
Wood Sap
♀2
Tm
☿2
Yb
☽2
Lu
W
S
Life Sap

Another example that stands out is Powdery Mildew on Grapes and Facial eczema in sheep, which are both fungal diseases that both respond to zinc applications. These are not the elements sitting at the February spot, but Zinc does sit opposite to this spot, thus suggesting we also explore the polar opposite 'companion' of the elements directly related to the seasonal phase. This exploration continues. Tb is suggested.

I will leave my initial suggestions here as 'thinking examples' and add the new options in italics. Using the trace elements associated with these regions provides the opportunity to emphasis specific bodies, by using a element from a particular Cosmic ring.

In the following suggestions there are the Biodynamic preparations in the suggestions. Here is a guide to their energetic activity.

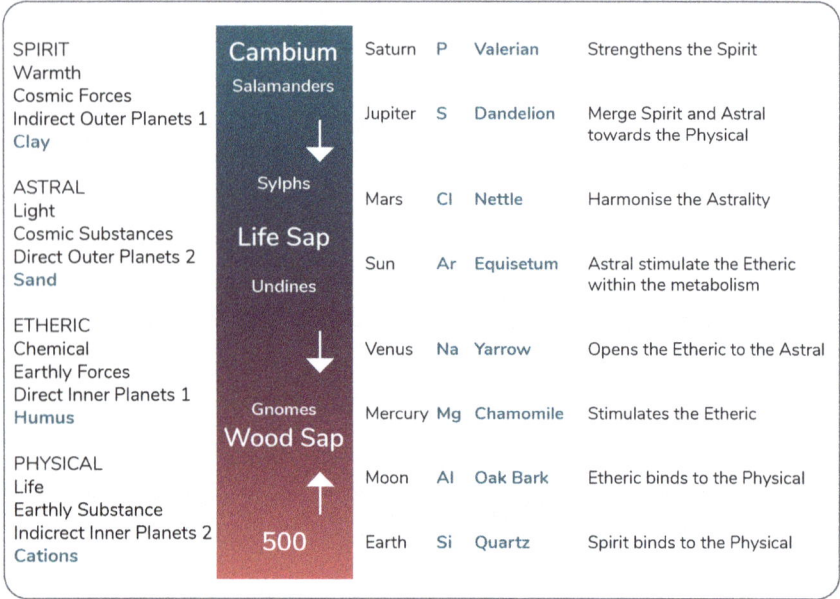

SPIRIT Warmth Cosmic Forces Indirect Outer Planets 1 Clay	**Cambium** Salamanders ↓	Saturn	P	Valerian	Strengthens the Spirit
		Jupiter	S	Dandelion	Merge Spirit and Astral towards the Physical
ASTRAL Light Cosmic Substances Direct Outer Planets 2 Sand	Sylphs **Life Sap** Undines	Mars	Cl	Nettle	Harmonise the Astrality
		Sun	Ar	Equisetum	Astral stimulate the Etheric within the metabolism
ETHERIC Chemical Earthly Forces Direct Inner Planets 1 Humus	↓ Gnomes **Wood Sap**	Venus	Na	Yarrow	Opens the Etheric to the Astral
		Mercury	Mg	Chamomile	Stimulates the Etheric
PHYSICAL Life Earthly Substance Indirect Inner Planets 2 Cations	↑ **500**	Moon	Al	Oak Bark	Etheric binds to the Physical
		Earth	Si	Quartz	Spirit binds to the Physical

Station 1 – Mid Winter to one month before Spring;
Earthly Substance phase; 2nd stage of the Life Ether / Wood Sap phase;
Aluminium / Boron

The crystallisation period occurs during this period. This is 'the' real centre point of the year.

One of RS stories talks of how the real 'seed fertilisation', where the Cosmic imprint is fructified for the coming season, does not take place until mid winter. So this is the beginning of the new season. The movement outward begins. The activities of the Earth, the Cosmic forces and the Earthly Substance have been stimulated by the Gnomes. The Cosmic Forces want to take the 'seed fertilised' species intention through to the new seed. The Earthly Substance wants to provide all the physical nutrition needed for the 'body' of the plant, making fine tissues. RS says it is the Gnomes revulsion of this physical activity that causes the spring to push out of the Earth. I am happy for this to be a natural process connected with the cycle of the Sun. The Silica Cosmic Forces want to shoot for the sky, and the Calcium Earthly Substance wants to hold on for the ride. This combination is aided through the elements of this period, Aluminium and it brother Boron. The Aluminium loosens the Silica from its structured lattice and as Clay, stimulates the Silica Cosmic Forces upwards. Boron helps, by facilitating Calcium to hold on for the ride. On this journey outwards the active Life Ether comes to an end, and the Etheric sourced Chemical Ether becomes more active. This is the journey of the Wood Sap.

500 is an obvious choice, due to it being in the Earth during this period, so it carries the Gnomes Calcium intentions.

I used 501 capped with Clay, this time, however I would just as happily use Clay by itself or a Winter Horn Clay may prove even better. The 501 is well used just as the first real growth begins. The closer you are to Spring when you apply this 'tonic', the more comfortably you can add 501.

I used Boron Phosphate, mostly due to a mix of elements from different layers appears to be more 'stimulatory', than a mix of elements of the same layer. I also like the stimulation of Phosphorus in the Spring to add with the light processes. Nitrogen could be a consideration as Boron's mate as it also need a boot along, once the soil warms up.

Regarding potencies, we are wanting the activity of the Earth to move above the surface. So middle potencies are OK now.

This application was applied in August, or a month or so after mid winter. It caused the plants to feel energised, however they appear like a cohesive mass of growth, jumping out to meet us.

This also suggests the group of elements around Cerium, Gold, Silver etc

Station 2 – Spring. One month before and after the Equinox; Chemical Ether is strengthening and the Light Ether begins; Life Sap begins; Magnesium.

Leaf growth is beginning, so photosynthesis is building the plant. The Etheric is growing in strength, and all the potential of the coming season lies ahead. Towards the end of this period the Astral inspired Light Ether begins to strengthen and the beginning of flowering processes begin.

This is one of the periods of the year when the Earth crosses the Sun's horizontal plane, which during the 'Lanthanide Journey' is experienced as a very nebulous ' in between' period where 'things are reconsidered'. In the Alchemical form, this is the Mercury time, which is the blending place where the poles meet. Magnesium Sulphate are the elements of the horizontal plane. Sulphur seems very appropriate with its catalytic abilities helping many elements and processes working together. Magnesium being the element sitting in the middle of the photosynthesis process seems also appropriate. It's energetic task is to bind the Etheric activity to the Physical, and this is mirrored in how the 'Life Sap' process which is beginning now, is bringing the Atmospheric Etheric activities to meet the rising 'Wood Sap' processes. This is the changeover period when the Below – the Earthly Substance process and the Above–The Earthly forces process have to interact.

So the 'middle' preps seem appropriate. We want to stimulate the Etheric and bind it with the Astrality, Chamomile, Yarrow, Stinging Nettle and Equisetum fit this need. Chamomile strengthens the Etheric, Yarrow opens it to the inward moving Astrality, Nettle ensures the Astrality is not too rampant, while Equisetum helps the Astrality and the Etheric to work together, bringing the Light into the Springtime moisture.

The chemical elements I used were Magnesium, Selenium, Nitrogen and Prometheum. Magnesium for its photosynthesis, Selenium, again to use a different layer to Magnesium, but also because this is the Etheric brother to Sulphur. The Nitrogen is to stimulate this element during the strong leaf phase, while Prometheum was added to hopefully bring some Spirit direction to the overall period, which can feel somewhat unfocused.

All potencies were in the middle range given we are still in the Etheric Leaf period.

This application was applied in late September or around the Equinox, and saw the raw growth forces of the Wood Sap become more refined and 'individualised', while the growth seems to stretch upwards with the increased Light forces.

It also suggests the group of elements around Barium Platinum and Palladium can be used especially if it is a wet spring.

Station 3 – One month after Spring to mid summer;
The Chemical and Light Ethers are active, while the Warmth Ether begins around a month or so before mid summer; Earthly Forces period; Life Sap builds in intensity so the Cambium process can build strength.

Growth is moving strongly, Light is stronger and towards the middle of this period the warmth processes are starting. My sense was that the garden was bursting out, with raw vitality, and needed some containing.

In October, or a month after Spring equinox, I applied a mixture of Chamomile, Yarrow, Nettle, Dandelion, Equisetum and Clay. The aim of this mix is to keep the Etheric and Astral working together, while bringing in the 'soft' Light and expansive processes of Dandelion into the fertilisation of the strong flowering of this period. The clay was added to keep the upward moving Cosmic Forces processes going.

The Chemical Elements were Potassium and Krypton. Potassium is used by the plant during this period to build strong structural stems and carry the Astral light processes into the Etheric, shown in the early fruit formation. Krypton was a interesting choice. I did not want to use the Halogen elements at this time of year, as they have a very condensing, even blocking influence of various processes. So I choose the Noble Gas group. This is a Internal Spirit influence, which should bring some internal warmth to the table. Krypton is the Etheric element of this group. Chosen in the hope that it would direct the Etheric to its proper jobs of growing leaves.

All potencies were in the low range to have the focus on the metabolic region of the plant and season dominate then.

This mixture was sprayed in October at the time that the Light Ether would be strengthening. The sense of the plants stretching out to the Light was reflected in the flower stalks of the time showing elongation between the flowers. The garden was crying out for some containment.

In November, at the time the Warmth Ether was beginning, and the Cambium process begins to move inwards, it was time for some 'contraction'. I decided to make up the Cosmic Substance mixture and apply it in November a month before Mid Summer.

The mixture contained Dandelion, Valerian, 501, Equisetum and Sand. The aim is to bring in the Spirit processes and stimulate the 'Cambium' mass formation.

The chemical elements I chose were Calcium Iodine. This mixture is meant more for after mid summer, but the beginnings of this period is when the Warmth Ether starts to function more strongly. Calcium is an element used by the plant during tissue formation and can be short as the fruit bulks up. It is also on the same ring as Potassium, which could probably be used instead now. I chose Iodine for similar reasons to Krypton. I did not want to use Chlorine, as indicated for the Cosmic Substance period, but did want to use a Halogen to bring this inward moving Spirit influence in. Iodine is the Astral element of this group. The Astrality stimulates things rather than shuts them off, and is found as the 'starter element' in many bio chemical processes, so I thought this would be the best choice, for this time of year.

All potencies were in the low series to focus things on the metabolic region.

This spray caused the plants to take on a more compressed and strong gesture. They stand strongly in their own space and begin to draw in substance and bulk up. Their colour went a deep green with a sense of strong vitality was present.

By mid summer – end of December – I felt the garden needed perking up. We have a had reasonable rain this season, so the usual drought stress has not been present yet. So I decided to do one application of Spray 2 and 3 together and then 2 days later repeat Spray 4. The Etheric stimulus had the plant more 'awake' but again they began to stretch to the Light. Spray 4 was being called for, even after 2 days, and after its application the 'stocky vitality' returned. The garden is a darker green than usual and things are growing strongly.

It also suggest the group of elements around Cesium Promethium Cobalt etc will be useful.

Station 4 – Mid Summer
I sprayed the No 2 and No 3 sprays together, in the morning. Two days later I sprayed the No 4 spray again.

Plants purked up and grew strongly. Very good weather 55mls of rain in January – mid summer.

It is suggested that the elements around Radon, Europium and Iron could be useful during this period.

**Station 5 – End of January to one week before Autumn equinox
Cambium should be strengthening, while the Cosmic Substance process begins to draw towards the Earth. The Chemical Ether moves to below the Earth.**

For the Cosmic Substance activity, I made a potion of 506, 507, 501, Sand and Equisetum at D3 focusing them into metabolic processes; Another 1/3 was 500, 502, 503 at D24 focusing them on the soil activity, while the last 1/3 was Palladium Chloride at D3. This is using the Sodium Chloride axis to highlight the Chlorine indication for this seasonal station. I liked the Palladium as it is a Astral stimulant of the cation arm, however being a transition element it works more into life processes. It is a 'horizontal' plane element. I am hoping it stimulates growth, for fruit sizing, while the chlorine is working to bring things to an end, in ripening.

This application seemed to create a severe contractive influence. The afternoon 501 influence appears to have over stimulated the high Phosphorus in my soil, causing a 'burnt out' effect.

This period is identified as when Powdery Mildew strikes grapes and cucurbits heavily, and facial eczema in sheep. Both fungal diseases that can be combatted with Zinc. Zinc sits opposite to this spot on the seasonal cycle. We also need to note RS suggested Mercury Sulphate for dry hot diseases. This is Zinc's bigger sister. So while it is suggested the elements around Atatine and Iodine such as Gd, Tb and Chromium are suggested, we have the hint that we should also look to the opposite elements that are suggested by the seasonal picture. I have proven using the Mercury and Zinc group has helped with Powdery Mildew, so this is an avenue worth pursuing.

Station 6 – Autumn Equinox
Life Ether becomes strong within the Earth,
Calcium, Magnesium Sulphate, 500, 505, Sand, D12 – 503, 506, 507,
508 D24

This spray had a gentle contracting influence, moving towards the Earth.

It is suggested that the group of elements around Pollonium Dy, Ta and Vanadium, can be useful during this period.

Station 7 – one month after Autumn till mid winter
Life Ether continues to build, while the Warmth Ether stops its above ground activity. This is the Cosmic Forces phase of the cycle. The Autumnal inward stream is joining with the Earth processes. The soil life is still active and Humus is being formed.

Etherics 1000 D24 – all BD Preps, Gallium Nitrate D24 to support these soil processes.

It is suggested that the elements around Bismuth Zirconium and Arsenic can be useful during this period.

Station 8 – Mid Winter – Crystallisation
The Gnomes are combining the Cosmic impulses received during the previous season, with the Earthly Substances available to them, to provide the 'fructification' and fertility activity for the coming season.

500, 501, 505, 507, Clay, Lime, Germanium Phosphate D 24

It is suggested that the elements around Lead Yb and Yittrium etc will be useful during this period.

Choosing elements from the specific Cosmic rings will help to specialise your actions. Spirit ring with bring warmth and order. The Astral ring will bring Light and activity to your processes, while Etheric elements will help when it is dry and more vitality is needed in the land.

Summary

Overall the 2019 season has been considered the 'best start to a two year drought', for the Hawkes Bay NZ. We had an even amount of rain and sunshine and warmth when we needed them the most, until we didn't.

My suggestion that the series of sprays mentioned here are all that would be needed, did not prove true. Fungal diseases, in tomatoes, were still evident and codlin moth in apples, came at the 'normal' untreated rate. Interestingly, powdery mildew was held in check by 'the season' responding when needed.

The overall garden health was great, due to the overall season. I will 'soften' the sprays through the summer period next year, as the 'inward' push was obviously too strong, for my high Phosphorus garden (600 ppm Melich 3 P – ideal is 50 ppm). If in doubt, I suggest staying with the 'Physical ring' chemical elements till greater experience is gathered of just how the outer rings influence plant growth.

In the following two seasons I have used only a few applications of my fungal complex – which is for both rots and dry fungus – with some Etherics 7 added, for pest control, and a dose of Etherics 1000. This has produced good overall results, when applied every 6 weeks or so.

Alchemical Chemistry

I see Biodynamic Chemistry and Alchemical Chemistry as two different reference systems, and try to not confuse the two. BD Chemistry is based upon the World Physical Arm and works well when thinking about the energetic bodies interactions. While Alchemical Chemistry, is based upon the Internal Physical Arm, and takes us into manifestation and to biological functions. Here the Physical Body organisation is divided into Nerve Sense (sal) Rhythmic (merc) and Metabolic (sulf) systems. Then we can consider which other body we might like to emphasis in that Physical region. This diagnostic method is the opposite of BD Chemistry, and appears to not need potentising.

An example was diarrhea, that arose from a mild poisoning by handling ant poison. The metabolic system was indicated however I specifically wanted to strengthen the Spirit activity there to bring a contracting order. Referring to the pictures on page 160, the metabolic region is the top of the chart, while the outside ring is the Spirit ring. We can see there is quite a lot of elements around that ring. I only use the inside of the two possible rings there, so far. This is focused upon the Internal Spirit whereas the outside ring of the Actinides, is more a Cosmic Spirit influence. I have not done much research into those yet. The Lanthanides though I have. When using this method I like to use one cation and one anion element, while there is something to be said for also including an element from the vertical axis, which appears to work as a 'director' of the overall Metabolic activity. So I could use Atatine (At) Radon (Rn) and Cesium (Cs). I could also choose to use the trace elements, or the lanthanides that are near these elements, or I could use all of them. I can also add the Sub Atomics that relate to the elements I chose, or just use the Sub Atomics by themselves.

In the event I used the Lanthanides Tb, Gd,Eu,Sm,Pm and the Sub Atomics Bos, Alp and Ire. 5 drops of each into my chocolate drink. Once. This 'pulled things together', and immediately stopped the possibility of 'accidents'. After a few days things were 'back to normal'.

Another example, is I have 'heart issues', having had 2 slight blockages in the veins on the outside of the heart. I also have high blood pressure. These are Rhythmic system issues and it became obvious that Magnesium and Selenium where suggested. The Magnesium increases the etheric into the Physical, while the Selenium is an Astral element that stimulates the Etheric, both in the rhythmic system. Once I took this I felt an immediate 'opening of my chest', where I could breath much more easily. I take them 'as needed'.

This approach is asking which energetic body do you wish to strengthen in which physical system. It appears no potentising is necessary.

These are the variety of ways I used my chemistry studies. There is no doubt more, so please explore any inspirations you might have. The good news is that by taking the remedies you are your own test subject, so you can see directly if they work as you imagine. Also plants are wonderful test subjects. They do not lie. Spray a plant and watch what happens, especially with the new growth, over the next two weeks.

Dosage

I suggest 'As Needed' is best. Homeopathics work best in small dosages, so we need to find what is the least we can use the remedy. The remedy sets a rhythm free into the environment, and it works for subtly and it can be for weeks. SO use it once and wait and see. Use 10 drops in a litre of water and spray to wet. 10 drops can be taken directly.

Walking the Path of Creation

The Manifest Journey

On page 138 is the diagram of how the planets and constellations relate to the Periodic Table. The planets order shown there and in this diagram, are an archetypal pattern. The astrological planetary rulerships for the zodiac sit like this, and are described by Dr Lievegoed within the context of the incarnation and excarnation processes of creation. The cycle begins with the Star / Spirit impulse entering into the planetary realm from the Cosmic / Galactic sphere, and being taken up by Saturn 1 as a 'divine task'. This the seed thought that is held firmly to, throughout the whole of the process of incarnation and manifestation. This is passed onto the progeny, be it a seed, your child or the inheritor of a social impulse. This seed has to go through many phases before the final end of one life cycle. Lievegoed's diagram shows this clockwise process (see 12, p. 237).

Once this patterning can be found, within a circle, and the periodic table, we can walk through the 'Process of Creation'. A meditative walk can be undertaken were you move around the circle starting at the Saturn 1 point. Have an intention you would like to process, if you wish, or just see what the process brings to you, as y ou move through it.

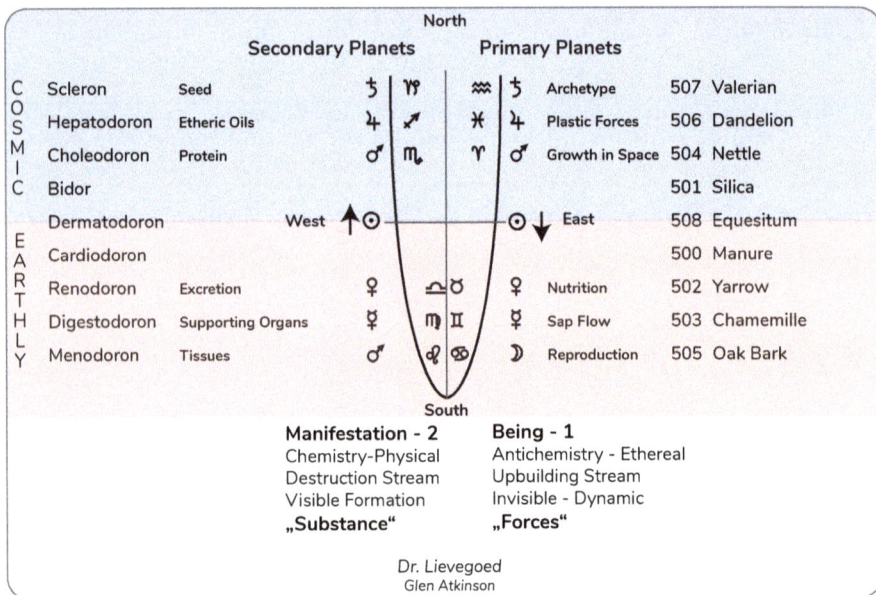

North

	Secondary Planets				Primary Planets		
C **O** **S** **M** **I** **C**	Scleron	Seed	♄	♑ ‖ ♒	♄	Archetype	507 Valerian
	Hepatodoron	Etheric Oils	♃	♐ ‖ ♓	♃	Plastic Forces	506 Dandelion
	Choleodoron	Protein	♂	♏ ‖ ♈	♂	Growth in Space	504 Nettle
	Bidor						501 Silica
E **A** **R** **T** **H** **L** **Y**	Dermatodoron		West ↑☉	‖	☉↓	East	508 Equesitum
	Cardiodoron						500 Manure
	Renodoron	Excretion	♀	♎☊ ‖	♀	Nutrition	502 Yarrow
	Digestodoron	Supporting Organs	☿	♍ ♊	☿	Sap Flow	503 Chamemille
	Menodoron	Tissues	♂	♌ ♋	☽	Reproduction	505 Oak Bark

South

Manifestation - 2
Chemistry-Physical
Destruction Stream
Visible Formation
„Substance"

Being - 1
Antichemistry - Ethereal
Upbuilding Stream
Invisible - Dynamic
„Forces"

Dr. Lievegoed
Glen Atkinson

In the same vein, you can do a series of homeopathic drop provings, that run you through the planetary sequence. Eg. Using the Transition Elements, you could start with taking a few drops of Iron / Fe. twice in one day. Do this for two days, as it takes half of the next day for the effect to change to the next one. On the third day take Cobalt / Co, twice, for two days. Then Nickel / Ni and so on through the series.

Observe how these drops effect you, and make notes of your experience and how your seed thought develops.

This process can be done with all the other layers and series as well.

The Archetypal Journey

A different pathway can be walked if you start at Moon 1 and walk anti-clockwise up to Saturn 1 and then back to Moon 2. This journey is the one taken when you walk along the 'atomic weight' increase of the elements. Jan Scholten uses this sequence for his understanding of the elements, and describes each layer of elements, as having an expanding and contracting process.

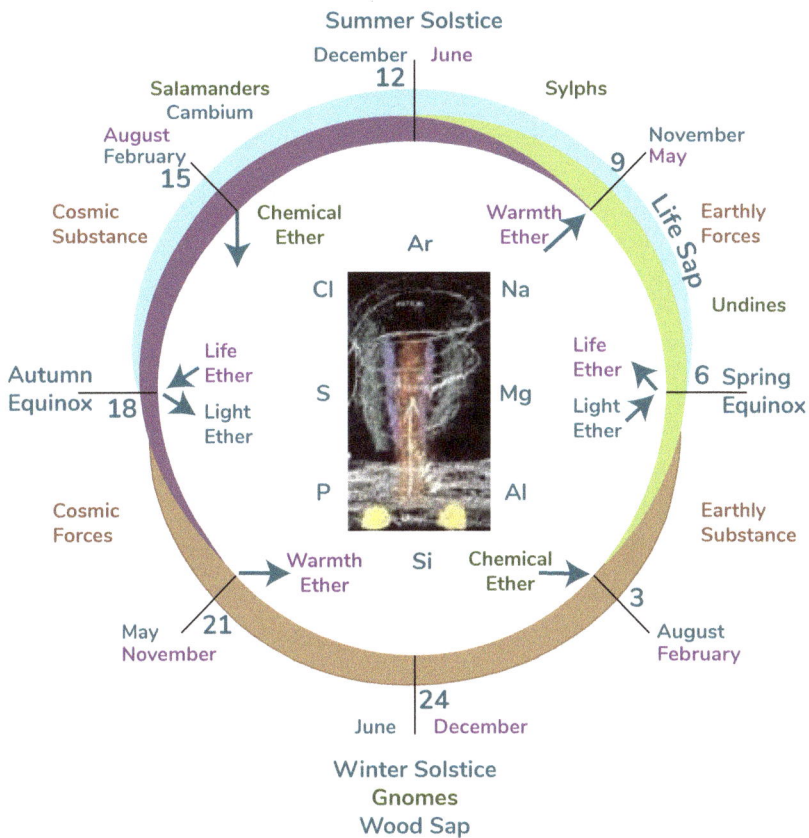

Summer Solstice
December | June
12

Salamanders
Cambium

August
February
15

Sylphs

November
May
9

Cosmic
Substance

Chemical
Ether

Warmth
Ether

Earthly
Forces

Life Sap

Ar

Cl   Na

Undines

Life
Ether

Life
Ether

Autumn
Equinox **18**

Light
Ether

S   Mg

Light
Ether

6 Spring
Equinox

P   Al

Cosmic
Forces

Earthly
Substance

Warmth
Ether

Si

Chemical
Ether

May
November
21

3

August
February

24

June | December

Winter Solstice
Gnomes
Wood Sap

This is a 'natural unfolding' sequence that can also be experienced through the Earth's Seasons. This is the Archetypal path. There is something very fundamental here. It is walking anti clockwise, and it can move your bodies around.

Here is the Seasonal Complex picture to help clarify the Seasonal activities. The planets journey (p. 126) and this picture are two different ends of the story. The Planets story is a Stage 1 story, while the Seasonal Complex is a Stage 3 story. Nevertheless they are both moving anticlockwise round the circle, just with different starting points.

Using 'Glen's Organising Device'

In the process of visualising the 3D spherical representation of periodic Table, I designed my clear and coloured 'wind sculptures'. These devices have been experienced to create a sense of calm and order in whatever environment they are hung in. Therapists in particular appreciate them hanging in their treatment rooms. They are designed to be blown by the wind, so they spin and also for the Sun to shine through them, thus throwing various types of light show about the room or garden.

They can also be used in a similar way to the paper image of the picture. One can hold the gyroscope and put your fingers upon a couple of points you wish to receive.

Alternatively, my experiments suggest that if you have a few small circular magnets, you can place two (so they stay in place) one on either side of a element, on the plastic disk, and leave it to hang in the environment you wish to influence. I am unsure just how far this will radiate, but I can say when I placed one of these within two meters of a sizable in ground wasps nest, that after two months the nest had stopped being active. In this season of huge wasp problems, we have had hardly any, after this.

Experimentation is the mother of invention. Go forth.

The Chartres Labyrinth

In Chartres Cathedral, one hour South of Paris, there is a sophisticated Labyrinth path, which has gained renewed popularity over recent times. The Chartres Cathedral was built during the 12th century by the Christian order of the Knights Templar. These knights had tremendous power during their time, and from their involvement in the crusades they had access to old wisdom, from both the European and more notably the Persian traditions.

The Labyrinth was developed as a replacement for pilgrims to walk, instead of making the journey to the holy land, which in the 13th century was no longer possible. It is a symbolic walk, from the physical life to the spiritual center of ones being. The walk begins from the South West end of the church and ends in the middle of the Labyrinth. One leaves the Labyrinth by walking directly up the Cathedral, towards the alter which is in the middle of the four arms of the cross, and onto the Holy of Holies, in the North East.

While many of the cathedrals where built on the North, South, East, West axis, with the main hall focused on the East West axis, Chartres is directed to the North East, so that the Sun rises in its North East window at the Summer Solstice, rather than the East, where the sun rises at the Spring Equinox. Such as Wells Cathedral in the UK.

There are many books, interpreting the architectural significance of the Chartres Cathedral, and I will leave it to them to provide the details, however there are few points of significance I would like to highlight.

In his book, "Chartres, Sacred Geometry, Sacred Space" Gordon Strachan, (from which this picture is taken) outlines many of the fea-

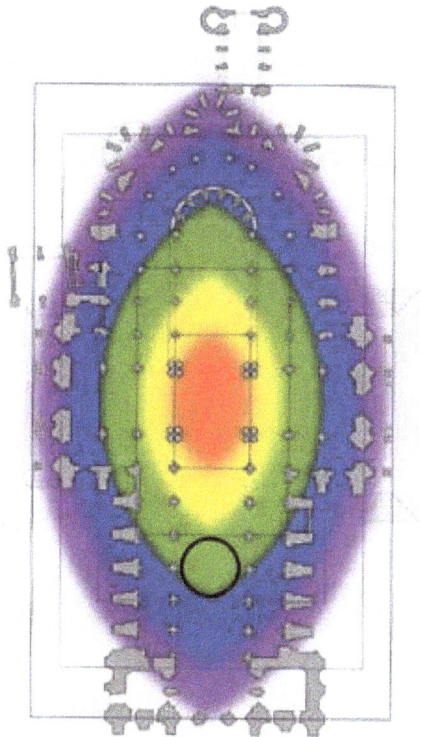

tures of the Cathedral. There is one feature which provides a indication of the significance the Labyrinth may play in the Cathedral. Strachan showed that the Christian symbol of the Viscsi Pisces was at the seat of the buildings design. He provided this picture, which shows the five Visci Pisces layers, he sees in the cathedrals design. It will be seen from this picture, that on the middle green ring, at top end, is the Holy of Holies. It was upon this focus, that all five of the previous churches built on this site, have been centered. This area is the spiritual focus of the whole cathedral. At the opposite or polar end of this green area is the position of the Labyrinth.

If the Holy of Holies is the spiritual center, and a place of inner experience of the 'mother and child', to whom all the sacred buildings on this site have been dedicated to, then the Labyrinth can represent, an exteriorised patterning of the archetypal form of creation. This being an inverted expression of the inner spiritual center.

Follow the path of the Labyrinth with your finger.

The Labyrinth conforms to the criteria I have identified in my other writings, regarding the structure of the Astronomical gyroscopic form standing behind creation, which in turn is mirrored in many octagonal sacred buildings and forms from every major culture over the last 10,000 years at least. This is a cross form, with twelve (6 x 2) internal layers.

I have a couple of unanswered questions though, 1) What direction is the appropriate entrance 2) How should the Periodic Table be aligned to the Labyrinth.

Labyrinth's are now being made all over the world, and there does not seem to be any agreement on where / which direction, the entrance should be. Chartres has its own quirk. Unlike the majority of Christian cathedrals and churches, which focus upon the East window, with their entrance from the West, Chartres is facing to the North East, to emphasis the position of the Sun at the Summer Solstice, with its entrance in the South West. Thus it would make sense to enter from the South West, however there have been many Labyrinth's in various cathedrals, and the Labyrinth always aligns with the main hall of the church, no matter which direction it is orientated too. As the majority, enter from the West then this might also be a more appropriate 'general' entrance point.

An interesting reference for the Labyrinth, may be the chapter houses built in the 1300s, especially the octagonal ones built in England, at Salisbury, Westminster Abbey and at Wells. All of these are orientated so that they are entered from the West. The Wells chapter house is particularly interesting, as above the outside row of seats are little orna-mental heads, representing who sat where. To the North sits the King, to the East sits the Bishop, to the South sits the leading Landowner. The North Apse is called the kings door and is where royalty would enter the cathedral, with the masses entering from the West. I wonder how much the design of the cathedrals built during the height of the churches era – pre 1500 – were designed to empha-sis the clergy and diminish the royalty?

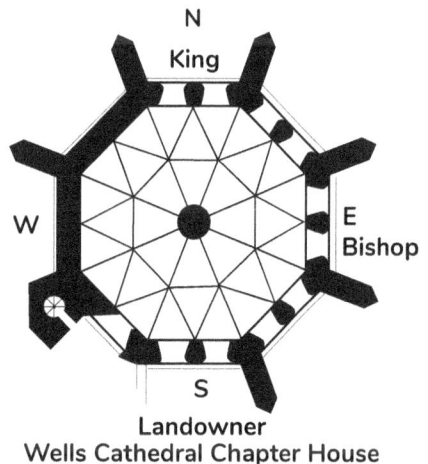

Wells Cathedral Chapter House

The Chapter Houses were where the business of the parish was carried out. In the case of the Westminster Chapter house, it served as the house of parliament for many years, before they moved to the present parliament buildings. So the chapter house is where 'stuff was sorted'.

Why the South West

With reference to my organisation of the energetic activities within the gyroscope, placed according to the archetypal electro–magnetic form, I have placed the World Spirit to the North, the World Etheric to the East, the World Physical to the South and the World Astral to the West.

'Trouble' comes from the Astrality. The astrality, is the name given to the energies coming to us as a result of the movement of the planets. It is therefore always changing, given the planets are constantly moving. As the planets move so our EM environment is influenced by their constant alteration, which in turn shows its influence upon human psychology. The Astrality is were all our issues are embedded. Hence the 5000 year old study of Astrology, which plots these planetary movements and subsequent human effects.

So if there are issues arising for the individual, or a community, then they will come largely from planetary movement and thus the Astrality. Therefore entering a place of resolution from the West, is symbolic of bringing in a problem, to be resolved by the community. Once it is addressed then the world can be re-entered with a 'clean astrality'.

For the Labyrinth we could apply a similar picture. Given it is considered a 'spiritual walk', with the aim being to 'sort stuff', so that we can be with spirit. The stuff to be sorted would be astral stuff. So entering from the West, into the Labyrinth would do the same thing.

At Chartres though it is aligned along the NE SW axis and which places the entrance on the internal physical / internal spirit axis, but firmly in the internal physical sphere of the EM form over the Cathedral. This axis personalises everything. The World arms are about things outside ourselves and thus correct for community related phenomena, however the internal arms make it our stuff. For the whole cathedral to have this orientation it would bring 'the issues' back to the personal and thus be a place that focuses the personal spiritual journey and what it means in ones daily life, more than when focused on the E W axis. Interestingly I believe it is the only remaining Labyrinth in French cathedrals, as all the others have been removed.

I have only had one short visit to Chartres, fairly early on in my octagon walking, with the main purpose being to identify that the energy changed as one moved about it, but as for identifying the specific energy of any one place it was still too early for me to define clearly. Also it was covered in chairs which made moving about difficult. My visit did show me I was on the right track though.

Initially, I overlayed my basic pattern, of the energetic bodies and the Periodic Table, over the labyrinth. This was published in my earlier edition of 'Biodynamic Chemistry'. However, I have since come to see a significant difference between the path I followed, based upon the 'Astrological orientation'. Which uses the Zodiac and thus the Sun's ecliptic, and the Zenith, as seen from the northern hemisphere; and the orientation based upon the Earths, North South electro magnetic field. I have concluded that as the Periodic Table is electro magnetic in nature, it should be organised according to the Earth's North South electro magnetic field, rather than the ecliptic/Zenith orientation, and the Labyrinth as found at Chartres, is a very specific Christian phenomena, and for very specific reasons is situated on the NE SW axis. This should be honored, so now I present the image of the Labyrinth along the directional line of the Chartres cathedral (NE / SW) with the Periodic Table to the North.

Inside the Labyrinth

The Labyrinth is 2D gyroscopic form, having a strong emphasis on the vertical and horizontal planes. It has four quadrants and 12 rings. These 12 internal rings, are identified in my 'Apple of Life' picture, and are formed by the 2 fold nature of the 6 major dimensional rings, described in 'Biodynamics Decoded'. It is therefore possible to (a) color code for the Labyrinth, (b) place my interpretation of the spiritual bodies interactions onto it. In this regard I prefer to follow the Periodic Table for guidance.

One further piece of information that is added to the overall collection of references, coming from the Labyrinth, is the relationship of the planets to each of the 12 internal layers. I have used my various sources of astrological reference to place the planets. This has the Earth at the center, with the other planets listed according to the order of their length of astronomical cycle, as experienced from the Earth. The last planet in the series is Persephone. I have talked of its influence in my "The Planets". This planetary order was confirmed as being 'correct', from documents cited from a Chartres lecturer, in Europe. This provides one more significant reference point to investigate all other accumulated information.

This pattern and form of spiritual antiquity, can be expected to have an archetypal power associated with it. With the addition of the Periodic table of elements to the Labyrinth, it could be possible this has identified the quality of various stages of the journey, around the Labyrinths 'Journey of Life'. It may also be possible to 'attract' any particular element, by standing on the appropriate place on the Labyrinth. Place your finger on a particular element and see if you can define their influence.

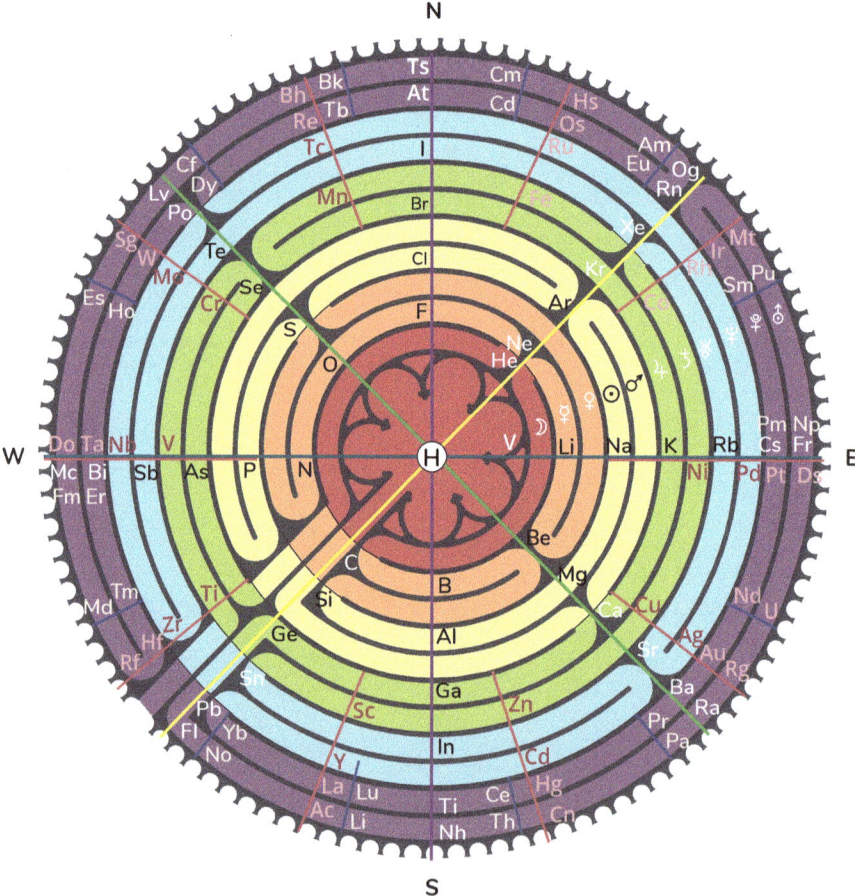

Appendix 1
The Atkinson Conjecture

In this modern world it seems everything needs a name. Within the scientific worldview there is this wonderful 'place' called the conjecture. This is where we are allowed to dream and makeup whatever story we wish, as a possible explanation of whatever we are interested in. Unlike the theorem, which has to be proven right in all circumstances, the conjecture is a beginning, something that is on the way to being proven. This is the process I have been involved in. I had a vision of what could be – in the late 70s – and have spent the intervening years showing how this 'idea' has truth.

The 'idea' was that there is an inherent archetypal order within living creation, and if we can identify this order we will be able to control life processes, in a manner that is ecologically safe and sustainable. This order can be identified as an inner expression of the outer structures we find about us. The next part of this conjecture is, that the processes identified, can be controlled within all living beings, by the Biodynamic preparations.

This in itself is not a new idea. 'As Above, So Below' is a very old concept. Where I moved this on, was to firstly identify that Dr Steiner's suggestions, and the insights of Astrology can be matched up, via the vortex form and then the gyroscope. This provides an organisational pathway by which Dr Steiner's very complicated indications can be rationally understood, which in turn allows for very practical tasks to be identified. Several of these outcomes have scientifically credible trials attached to them. The most significant being the Bird control trials carried out by HortResearch NZ (12). This is a homeopathic product made from the BD preparations, that addresses the energetic discordance that draws a predator to its 'prey'. In the same manner that a sound can be cancelled by playing its opposite, so the energetic copy of an animal can cancel out its ability to be in a particular place. Identifying the energetic makeup of any being becomes the task, and one the biodynamic community has made great strides towards.

For a conjecture to be true it must be right in all circumstances, which meant that chemistry – a organic manifestation of our environment – must also conform to 'the archetypal order'. This publication and the practical trials that have developed from it, provides the proof of this. My efforts to date prove to me the Atkinson Conjecture is indeed a truth, and thus its formulas, **Biodynamic Vortex, Biodynamic Glossary, Energetic Bodies Interaction, 3D Periodic Table, Glenological Rosetta Stone, and Creation = Movement + Time** are a theorem.

Appendix 2
From Double Cross to Gyroscope

The Three Dimensions of Space

It is not very often in modern life that we consciously have to think in three dimensions. Naturally we 'live' in three dimensions, always making judgments based on distance, speed and our relationship to things, but when it comes to our thinking, we so easily take a lot of things for granted as being flat. When we look at a picture of our Galaxy, how often do we see the flat stationary spiral plane. How many of us add to this, that this is spinning at 64,000mph or that this flat plane we see, is just one part of a gigantic spherical ball?

While attempting to co-ordinate all the various pieces of information we find in Biodynamics, I began with a flat piece of paper, upon which I eventually came to the form of a flat vortex, as the best form for the organisation of the multi layered picture, given to us by Dr Steiner. With some more 'perceiving' it became clear that vortexes only exist within gyroscopic spheres. Thus I moved with my flat piece of paper, to the 8 segmented, six ring diagram I have used for the "Agriculture Course Glossary". This sufficed for some time until my investigations into the Periodic Table led me to seeing the atomic structure of any chemical element was a 3 dimensional sphere. This led to the question as to how to make the Periodic Table into a 3 dimensional form. At this point I moved from the flat planes, with no depth, to a significantly more complex 'being' with all sorts of new relationships to consider. I initially made paper and cardboard 3D models, before moving to laser cut plastic models, which have allowed for some sturdiness, with colour elements being included,

as well as allowing for the impact of wind, to add the necessary element of movement.

The main features of a 3 dimensional model are three planes. One indicating height, one for depth and one for width. However this is not a stationary being. It is a being created out of movement. Spinning gyroscopes, initially form vortexes on the vertical axis, through which they draw in substance to their centre, before squirting this out again along the horizontal plane, where it often accumulates, as the substance we see as planets or even the flat spiral form of the galaxy.

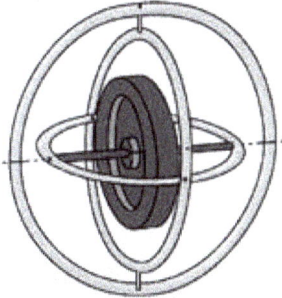

This identifies that there are two vertical axis and one horizontal plane. So we can see that the vertical axis do represent 'upwards', however it is probably more appropriate to talk of north and south or positive pole and negative pole, based upon the electro magnetic fields these 'beings' live within and organise accordingly. A further distinction to this form came, when this form was drawn according to the arrangement of the chemical elements, and how I allotted a particular group of elements to each plane. (see Glenopathic Chemistry). This identified the primary vertical with the major elements, the secondary vertical with the Rare Earth elements, which only manifest on the outer ring of this plane, and the transition elements on the horizontal plane. These only begin to manifest at the fourth ring of elements, presenting a horizontal ring similar to that seen around the planet, Saturn. These differences help to see each plane as having quite different qualities.

The exact nature of the difference – other than their relationship to the different groups of chemical elements and their 4th, 5th, & 7th 'tones' – did not concern me much until I noticed the significance of the difference of orientation, between our cultural focus on the Sun and Zodiac, and what derived from orientating on the north magnetic pole of the Earth. I had in some way taken these to be the same. This 'smudging' is made easier as I live in the southern hemisphere, and so I naturally orientate to the north, when I look towards the Sun and the Zodiac. In 'Two Orientations' I explore this question further.

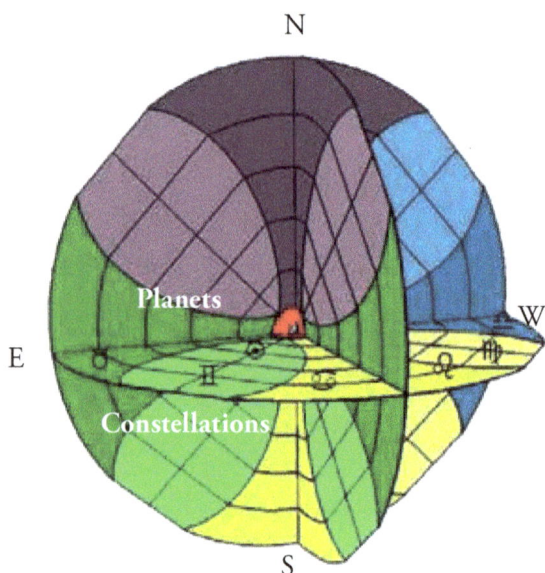

The key point for this discussion is that the planets all exist on the horizontal plane of the Sun gyroscope. When we look from the Earth, at the Sun, we see it amongst the other planets, along the line we call the ecliptic. Depending where we are (what latitude) on the planet, this arc across the sky will be either higher or lower in the sky. However if we are in the northern hemisphere we will look towards the South to see the Sun's path, which means the east is on our left hand and the west is on our right hand.

The Zodiac is defined as those groups of stars, standing behind the path of the Sun. Therefore the zodiac is also a being of the horizontal plane. This orientation to the Zodiac / Sun is a very old practice within our culture, and we can see that Dr Steiner naturally used the Zodiac and planets as major references in his work.

As juxtaposition to this orientation, we need to consider the possibility of orientating off the north magnetic pole. Some ancient cultures used this Earth reference, to place their temples and cathedrals, however it seems most were orientated to the Sun. Considering we are electro magnetic beings, there is something to be said for us to align our personal north south poles, more consciously to the north south poles of the Earth...

In my wanderings as to what the different orientations might mean, I came across a few passages by Dr Steiner in his 1920 lecture series "Man–Hiero-glyph of the Universe" One of Steiner's central tenets was that if we are ever to understand ourselves we must first appreciate the order of the Universe about us, and then see how we are imaged upon this 'master plan'. Alternatively of course we can find things in ourselves and see how these are reflected outside ourselves.

In lectures 1 and 3 of this series of lectures he talks of the importance of living into the three dimensional planes. He says "Astronomy observes the courses of the stars and calculates; but it notices only those forces which show the Universe, in so far as the Earth is enclosed in it, as a great machine, a great mechanism. It is true to say that this mechanical-mathematical method of observation has come to be regarded simply and solely as the only one that can actually lead to knowledge.

Now with what does the mentality which finds expression in this mathe-matical-mechanical construction of the Universe reckon? It reckons with something that is founded to some extent in the nature of Man, but only in a very small part of him. It reckons first with the abstract three dimensions of space. Astronomy reckons with the abstract three dimensions of space; it distinguishes one dimension, a second (drawing on blackboard) and a third, at right angles. It fixes attention on a star in movement, or on the position of a star, by looking at these three dimensions of space. Now man would be unable to speak of three dimensional space if he had not experienced it in his own being. Man experiences three-dimensional space. In the course of his life he experiences first the vertical dimension. As a child he crawls, and then he raises himself upright and experiences thereby the vertical dimension. It would not be possible for man to speak of the vertical dimension if he did not experience it. To think that he could find anything in the Universe other than he finds in himself would be an illusion. Man finds this vertical dimen-sion only by experiencing it himself. By stretching out our hands and arms at right angles to the vertical we obtain the second dimension. In what we experience when breathing or speaking, in the inhaling and exhaling of the air, or in what we experience when we eat, when the food in the body moves from front to back, we experience the third dimension. Only because man experiences these three dimensions within him does he project them into external space. Man can find absolutely nothing in the Universe unless he finds it first in himself. The strange thing is that in this age of abstractions which began in the middle of the fifteenth century, Man has made these three dimensions homogeneous. That is, he has simply left out of his thought the concrete distinction between them. He has left out what makes the three dimensions different to him. If he were to give his real human experience, he would say : My perpendicular line, my operative line, my extensive or extending line. He would have to assume a difference in quality between the three spatial dimensions. Were he to do this, he would no longer be able to conceive of an astronomical cosmogony in the present abstract way. He would obtain a less purely intellectual cosmic picture. For this however he would have to experience in a more concrete way his own relationship to the three dimensions. Today he has no such experience. He does not experience for instance the assuming of the upright position, the being in the vertical; and so he is not aware that he is in a vertical position for the simple reason that he moves together with the Earth in a certain direction which adheres to the vertical. Neither does he know that he makes his breathing movements, his digestive and eating movements as well as other movements, in a direction through which the Earth also moves in a certain line. All this adherence to certain directions of movement implies an adaptation, a fitting into the move-ments of the Universe. Today man takes no account whatever of this concrete understanding of the dimensions; hence he cannot define his position in the great cosmic process. He does not know how he stands in it, nor that he is as it were a part and member of it. Steps will have now to be taken whereby man can obtain a knowledge of Man, a self-knowledge, and so a knowledge of how he is placed in the Universe.

The three dimensions have really become so abstract for man that he would find it extremely difficult to train himself to feel, that by living in them, he is taking part in certain movements of the Earth and the planetary system. A spiritual-scientific method of thought however can be applied to our knowl-edge of Man. Let us therefore begin by seeking for a right understanding of the three dimensions. It is difficult to attain; but we shall more easily raise ourselves to this spatial knowledge of Man if we consider, not the three lines of space standing at right angles, but three level (equal) planes. Consider for a moment the following. We shall readily perceive that our symmetry has something to do with our thinking. If we observe, we shall discover an elementary natural gesture that we make if we wish to express decisive thinking in dumb show. When we place the finger on the nose and move through this plane here (a drawing is made), we are moving through the vertical symmetry plane which divides us into a left and a right Man. This plane passing through the nose and through the whole body, is the plane of symmetry, and is that of which one can become conscious as having to do with all the discriminating that goes on within us, all the thinking and judging that discriminates and divides. Starting from this elementary gesture, it is actu¬ally possible to become aware of how in all one's functions as Man one has to do with this plane,

Consider the function of seeing. We see with two eyes, in such a way that the lines of vision intersect. We see a point with two eyes; but we see it as one point because the lines of sight cross each other, they cut as shown in the drawing. Our human activity is from many aspects so regulated that we can only understand its regulation by reference to this plane.

We can then turn to another plane which would pass through the heart and divide man back from front. In front, man is physiognomically organised, behind he is an expression of his organic being. This physiognomical-psychic structure is divided off by a plane which stands at right angles to the first. As our right and left man are divided by a plane, so too are our front and back man. We need only stretch out our arms, our hands, directing the physiognomical part of the hand (in contrast to the merely organic part) forwards and the organic part of the hands backwards, and then imagine a plane through the principal lines which thus arise, and we obtain the plane I mean. In like manner we can place a third plane which would mark off all that is contained in head and countenance from what is organised below into body and limbs. Thus we should obtain a third plane which again is at right angles to the other two.

One can acquire a feeling for these three planes. How the feeling for the **first** (vertical) is obtained has already been shown; it is to be felt as the plane of discriminative *Thinking*. The **second (vertical)** plane, which divides man into front and back (anterior and posterior) would be precisely that whereby man is shown to be Man, for this plane cannot be delineated in the same way in the animal. The symmetry plane can be drawn in the animal but not the vertical plane. This second (vertical) plane would be connected with everything pertaining to human **Will**. The third, the horizontal, would be connected with everything pertaining to human Feeling. Let us try once more to get an elementary idea of these things and we shall see that we can arrive at something by this line of thought.

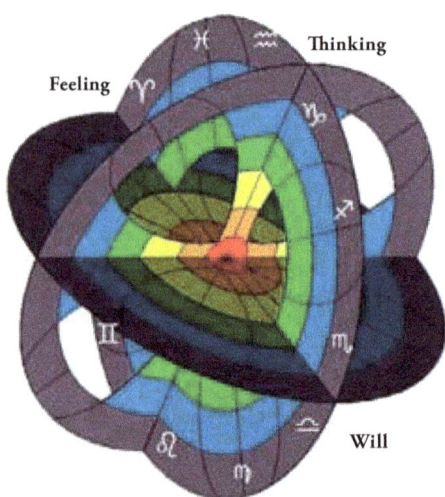

Everything wherein man brings his feeling to expression, whether it be a feeling of greeting or one of thankfulness or any other form of **sympathetic feeling**, is in a way connected with the **horizontal plane**. So too we can see that in a sense the will must be brought into connection with the vertical plane mentioned. It is possible to acquire a feeling for these three planes. If a man has done this, he will be obliged to form his conception of the Universe in the sense of these three planes – just as he would, if he only regarded the three dimensions of space in an abstract way, be obliged to calculate in the mechanical-mathematical way in which Galileo or Copernicus calculated the movements and regulations in the Universe. Concrete relations will now appear to him in this Universe. He will no longer merely calculate according to the, three dimensions of space; but when he has learnt to feel these three planes, he will notice that there is a difference between right and left, over and under, back and front. In mathematics it is a matter of indifference whether some object is a little further right or left, or before or behind. If we simply measure, we measure below or above, we measure right or left or we measure forward or backward. In whatever position three meters is set, it remains three meters. At most we distinguish, in order to pass from position to movement, the dimensions at right angles to one another. This we do, however, only because we cannot remain at simple measure¬ment, for then our world would shrink to no more than a straight line. If however, we learn to describe Thinking, Feeling and Willing concretely in these three planes, and to place ourselves thus in space as psychic-spiritual beings, with our Thinking, Feeling and Willing then just as we learn to apply to Astronomy the three dimensions of space as found in man, so do we learn to apply to Astronomy the threefold division of man as a being of soul and spirit".

In lecture 3 the plans are further clarified.

"When we are aware of ourselves as Man standing on the Earth, surrounded by the planets and fixed stars, we begin to feel ourselves as part of all these; it is not a matter merely of drawing three dimensions at right angles, but of thinking concretely about the Cosmos and penetrating into the concrete reality of the dimensions.

Now there is a series of constellations that is immediately evident to those who study the outer Universe at night-time, and has indeed always been seen when men have studied the stars. It is what we call the Zodiac. It is immaterial whether we believe in the Ptolemaic or the Copernican system; if we follow the apparent course of the Sun it always seems to pass through the Zodiac in its yearly round. Now if we imagine ourselves placed into the Universe in a living way, we find that the Zodiac is of very great significance. We can¬not conceive of any other plane in celestial space as being of like value with the Zodiac, any more than we could conceive the plane which divides us in two and creates our symmetry, as being placed at random just anywhere. We then perceive the Zodiac as something through which a plane may be described. Let us suppose this plane to be the plane of the blackboard, so that we have here the plane of the Zodiac; the plane of the Zodiac is just the plane of the blackboard. We shall then have one plane before us in Cosmic space, precisely as we imagined the three planes sketched in Man. That is certainly a plane of which we can say that it is fixed there for us. We see the Sun run its course through the Zodiac; we relate all the phenomena of the heavens to this plane. And we have here an analogy of an extra-human kind for what we must perceive and experience as planes in Man himself. Now when we draw the Symmetry plane in Man, and have on one side of the Symmetry-axis the liver organised in one way, and on the other side the stomach organised in a different way, we cannot think of such a fact without feeling at the same time some inner concrete relation; we cannot imagine mere lines of space lying there, but what is in the space must manifest definite forces of activity; it will not be a matter of indifference whether something is on the right or on the left. In the same way we must imagine that in the organisation of the

Universe it is a matter of consequence whether a thing is above or below the Zodiac. We shall begin to think of Cosmic space – as we see it there, sown with stars – we shall begin to think of it as having form.

Now just as we can think of this plane on the blackboard, so we can also think of another at right angles to it. Let us think of a plane extending from the constellation Leo to that of Aquarius on the other side. Then we can go further and imagine a third plane at right angles again to this one, running from Taurus to Scorpio. We have now three planes at right angles to one another in Cosmic space.

These three planes are analogous to the three we have imagined described in Man. If we think of the plane we have denoted as that of Will – the plane namely which separates us behind and before we have the plane of the Zodiac itself.

If we think of the plane running from Taurus to Scorpio, we have the plane of Thinking; that is, our Thought plane would be co-ordinated to this plane. And the third plane would be that of Feeling. Thus we have divided Cosmic space by means of three planes, just as we divided Man in our first lecture. What is primarily of importance is not simply to unlearn as quickly as possible the Copernican Cosmic system, but to enter into this concrete picture, to imagine Cosmic space itself so organised that one can distinguish in it three planes at right angles to one another, just as can be done in the case of Man."

While quoting RS there is a relevant passage in lecture 12, of the 1920 medical lectures, regarding some substances relationship to the three directions. "You know that certain substances operate in the human organism simply through being bound up with either bases or acids; or appear, to use the technical term, neutrally in the form of salts. Thus bases and acids act as complexes of antagonistic forces, which neutralise each other in salts. But this is not all. How does this triad, acids, bases and salts, operate within the human system of organic forces? We shall find that all bases have a tendency to support such human processes as begin in the mouth and continue through digestion, i.e., from front to rear; and indeed all other processes with the same line of action. And as the basic substances have to do with this direction, so the acids are equally associated with the reverse. Only in studying the opposition of "front man" to "rear man" one understands the polar opposition of bases and acids. And saline substances stand at right angles to the two opposites, pointing vertically earthwards. All processes directed from above downwards centripe-tally are those into which the saline element thrusts itself. Thus we must keep these three spatial directions clearly in our minds, if we seek to determine how man enters into the triad, bases, salts and acids."

While the allotting of the activities of Thinking, Feeling and Willing to the three planes of the Gyroscope are significant references, the most important reference RS makes here is that the Will plane is the plane of the Zodiac. He earlier described this as the second vertical plane. From this we can make up the three dimensional diagram of his vision.

However, with his picture comes some interesting 'difficulties'. If we just take RS words and have this nice picture and look no further we can rest in some degree of peace, however once we take his next suggest and look into the nature and activity of all these spaces created within this 3 dimensional sphere things become somewhat more complex.

The first issue that arises is, that RS image does not correspond with the Astronomical reality of the Sun's gyroscope – the Solar System, as the Zodiac is a being of the horizontal plane of the Sun gyroscope, and not a vertical axis. The Zodiac is identified as those star groups behind the Suns path – the ecliptic. Remembering we are looking from the Earth, so the Sun takes the place of the Earth from our perspective, thus the Sun/Earth and the planets manifest on the Sun's horizontal axis, thus the Zodiac also 'lives' on the Sun's horizontal plane. This is an image of the issue I raised at the beginning of this article. Here RS is taking the archetypal horizontal plane and 'making it' a vertical plane.

Many questions arise out of this: How do these two 'truth's relate to each other? Why would RS do this and what does it mean when it is done? There are also several issues of orientation of the gyroscope. Which side is Up and where is East and West.

My answers to these questions are given as an expression of the spirit of RS comments "When we are aware of ourselves as Man standing on the Earth, surrounded by the planets and fixed stars, we begin to feel ourselves as part of all these; it is not a matter merely of drawing three dimensions at right angles, but of thinking concretely about the Cosmos and penetrating into the concrete reality of the dimensions."

Organising the Three Dimensions

We need to go through a process of building up references so we have the base for assessing what it means for RS to put the Zodiac as a vertical axis.

Step 1 – Universal order

Creation = Movement + Time

Once anything moves, it begins to spin which leads to the development of a electro magnetic field, which in turn orders all forces and matter in that field into the form of a complex 'cross' gyroscope. The primary structure of this form, has a vertical axis with two vortex forms, which draw in substance and forces to be consolidated in the center, before being squirted out along the horizontal plane.

In this ordering we have a north and a south pole, each drawing in matter from their pole. It is therefore a good question as to which way is up. If we are looking at a star in space then there is probably not much difference between the two, however once we apply this form to living beings on the Earth, such as plants or humans then the image becomes clearer. For life forms we have a duality of influences. Those coming from above in the form of light and warmth, and those coming from the Earth in the form of Water and Earth. Thus the poles can be differentiated into the Cosmic pole coming from above – North and the Earthly pole coming from below – South.

This axis can be considered to be generally fixed in life forms, for our present purposes. However this form is not stationary. It is spinning, and thus the question of where is East and where is West becomes an interesting question. Firstly we have to stop the spinning and freeze it for a moment, and then we have to consider where we are orientating ourselves. We have two choices. Lets take it that we are on the Earth, and so we can orientate off the North Pole, which is THE 'real' vertical axis OR we can orientate off the ecliptic, as seen from the northern hemisphere, which is a specific view of the horizontal axis. The significant question that arises is which side / hand is the East and West placed.

Most of my diagrams have been orientated off the ecliptic, from the northern hemisphere, and so as this is THE most unconsciously accepted orientation in our culture, given the Sun and Zodiac are used as our major references, I will use these for the moment.

When looking to the South to the ecliptic, we therefore have the East on our left hand side and the West on the right hand side.

Step 2 – Orientation

Upon this basic archetypal form we can develop the organisation of the energetic bodies in life, commonly called the Spirit, Astral, Etheric and Physical forms. This placement is based upon Dr Steiner's indication in his October 1922 medical lectures.

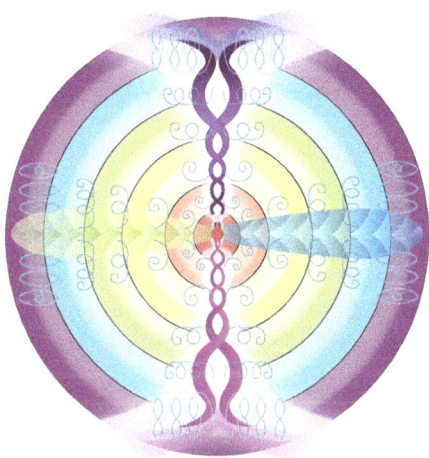

Here we have the organisation of the World activities upon the primary axis and the internal activities of the bodies on the secondary cross.

While this identifies the activities on a flat two dimensional image, the question is what happens to this when it is moved to a 3 dimensional form.

A basic feature of the three dimensional form is that there are three planes, Height Width and Depth.

From working with the Periodic Table of elements it is possible to gain some definition of each of these planes. I have placed the three groups of chemical elements upon the three different planes These are the Major elements, the Trace elements and the Rare Earth elements.

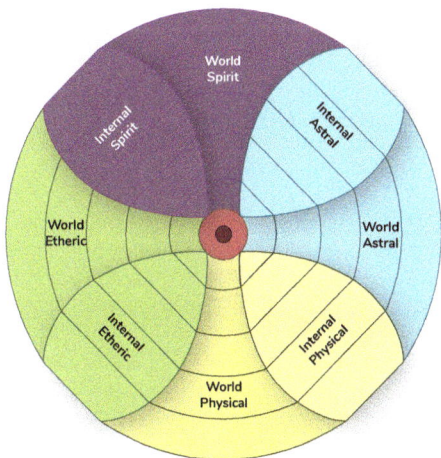

We can also place the planetary organisation over these three groups. This particular organisation comes from Biodynamics and particularly Dr Lievegoed. It identifies an 'inward' moving primary planetary activity and a 'outward' moving secondary planetary order. This is useful for identifying the 'top' of the gyroscope in relationship to the planets, as in biodynamics it is clear that Saturn is the carrier of the Spirit activity within the planetary realm, thus we can allocate Saturn to the top of the World Spirit, vertical axis. The planetary reference is useful when we come to orientating the zodiac, as the constellations have specific planetary ruler ships. Thus we can place the Zodiac with Aquarius and Capricorn at the top of the gyroscope. One question arises when allotting the planets to the zodiac and this is which constellations are related to the 'primary and which are the 'secondary' functions of the planets? I have addressed this question in my "12 fold Manifestation".

3D Plans

The next step is to put these three planes together to form a three dimensional gyroscope.

Step 1 – We firstly place the three layers on top of one another, while holding the purple vertical axis together and at the top.

Step 2 – You need to physically face to the Sun's highest point in the sky – the zenith point of the ecliptic, as seen in the northern hemisphere. Thus the green vortex is on your left hand.

Step 3 – Holding two layers at the top, lay down the (Trace Elements) horizontal plane 90 degrees, so as to make complete green (a) and blue (b) vortexes on the horizontal plane.

Keeping with the reference of the Zenith having Spirit pole qualities, the purple vortex of the horizontal plane is laid down away from you, in the direction of the zenith.

Step 4 – The secondary vertical (Rare Earth) plane has to be spun, to 90 degrees of the primary (Major elements) vertical plane. This makes the purple /north and yellow / south vertical vortexes complete. Again this spin can go either way. I chose to spin the blue vortex towards the zenith, keeping the two 'male' activities together.

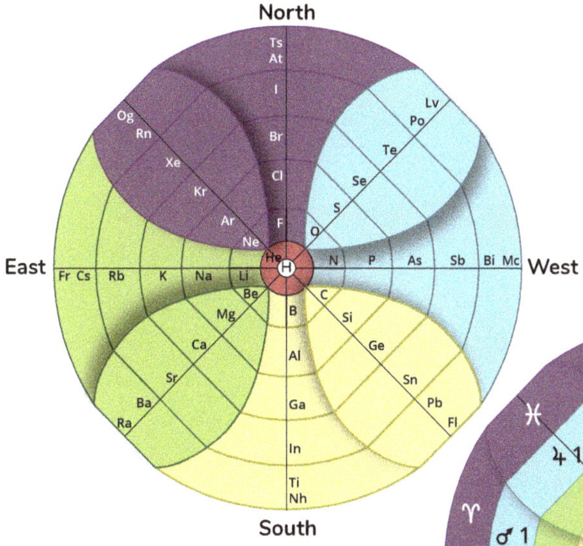

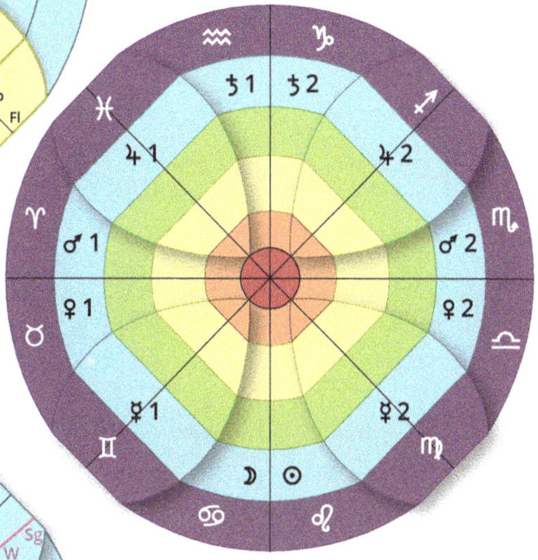

North

Ts
At
I
Br
Cl
F

Lv
Po
Te
Se
S
O

N P As Sb Bi Mc — West

East — Fr Cs Rb K Na Li Be

Og Rn Xe Kr Ar Ne He

Mg
Ca
Sr
Ba
Ra

B
C
Si
Al
Ge
Ga
Sn
Pb
Fl
In
Ti
Nh

South

Planets and Constellations

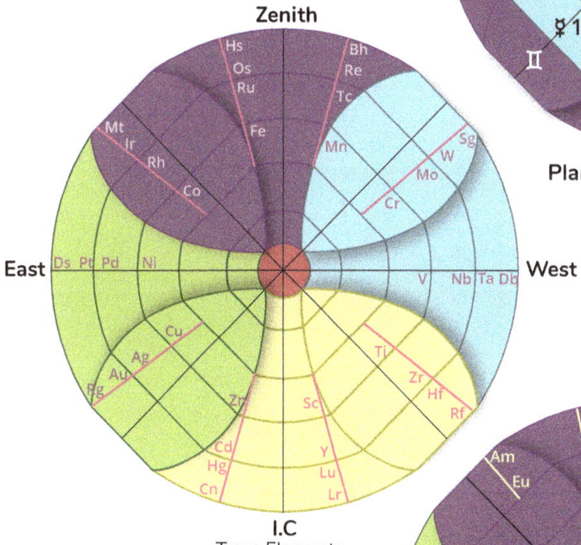

♒ ♑
♓ ♄1 ♄2 ♐
♈ ♃1 ♃2 ♏
♂1 ♂2
♉ ♀1 ♀2 ♎
☿1 ☿2 ♍
♊
☽ ☉
♋ ♌

Zenith

Hs
Os
Ru
Fe
Bh
Re
Tc

Mt
Ir
Rh
Co

Mn
Sg
W
Mo
Cr

East — Ds Pt Pd Ni

Cu
Ag
Au
Hg

V Nb Ta Db — West

Ti
Zr
Hf
Rf
Zn
Sc
Cd
Y
Hg
Lu
Cn
Lr

I.C
Trace Elements

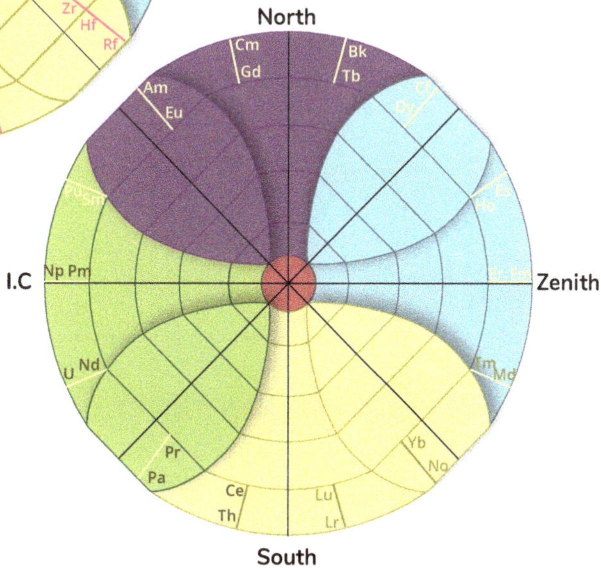

North

Cm
Gd
Bk
Tb

Am
Eu
Cf

Dy

I.C — Np Pm

Fm Md

Zenith

U Nd
Yb
No

Pr
Pa
Ce
Th
Lu
Lr

South
Rare Elements

This creates an interesting circumstance where the horizontal plane now has complete Green (a) and Blue (b) vortexes as part of the primary 'Majors' gyroscope. Then there are two vortexes of mixed colours, one (d) with purple and blue and the other (c) with yellow and green, where the Rare Elements plane crosses the Traces plane.

Translating this to the energetic body language, this gives an axis that has a spirit & astral qualities in a polarity to a physical & etheric pole.

This is the combination of activities we find in nature and living organisms. We have the Physical and etheric forces coming upward out of the Earth and the Spirit and Astral activities coming from above moving downwards.

Now we have a three dimensional form, we can note that the zodiac exists on the horizontal plane of the Solar gyroscope.

Dr Steiner's suggestion

In "Man–Hieroglyph of the Universe" RS describes the three planes of the gyroscope and he allocates the thinking feeling and willing processes to the three planes, with particular reference to the Zodiac. However he places the Zodiac as a vertical plane. Therefore this gives a specific orientation to the gyroscope, with the mixed vortexes (Purple Blue and Green Yellow) taking the vertical axis, while the axis of the major elements becomes the horizontal 'feeling' plane.

What I find interesting here is that RS is describing the relationship of the human to the gyroscope and thus he is using the orientation that places the arrangement of the bodies as we find them in nature as the basis of orientation.

RS talks of living into the spaces created by the divisions of the three planes and using this diagram it should be possible to identify the spiritual activity of each of the zones, however there is a problem in that RS did not define which way he was facing, with regards the zodiac. If we take it that he was continuing on with the northern hemisphere orientation, then the zodiac will be above and the east / green will be on the left hand side with the west / blue on the right hand side. Therefore the Back Right Below zone will have a mixture of Spirit Astral and Physical qualities. For the Front Right Above Zone– facing away from us – would be zone of some Astral and some Physical qualities.

I put these suggestions here just to show the process, as I would like to find other references from some other sources as a checking mechanism, before I go too far down this track.

Changing Orientation

Once the axis' have been designated to Thinking Feeling and willing, we can make a further observation about the significance of the change from the North pole orientation to the horizontal plane. The North Pole axis is the feeling axis, while the horizontal axis is the Will. Hence we have an image of humanity shifting from Feeling to Willing, with this change in orientation.

Thus the vertical axis of Thinking and Willing, over our feelings, can also be seen as an apt description of our 'modern' history.

Afterthought

While putting all this material together, and looking at the duality of orientation that has emerged between Astronomical realities and abstractions we humans have developed, (10) and then how I have framed this phenomena as a polarity of Cosmic and Earthly, or archetypal and subjective activity, and then looked at this duality of orientating to the Sun's ecliptic or to the Earth's north pole, which really means orientating to the Earth's magnetic field, I wonder….. I have also characterised the movement of humanities focus from the Constellations to the Signs, and thus a move from the Galaxy to our Sun, as a process of humanity internalising the Spirit. Instead of 'God the father' / Galaxy, being dominant and holding our fate in his hands, we have, by the end of the 20th century, internalised the Spirit and formed our own relationship with 'God', without reference to outside priesthoods and the like. This state is a fulfillment of the Piscean age challenge set down by the Christ. 2000 years he represented – Get one with God and be nice to people – which is now available to all of humanity.

Now I wonder how taking the next step, and orientating off the Earth's organisational field, rather than the Sun's path, we are in the process of 'coming to Earth' and 'killing God the father'. So the stance is one of me and the Universe, not so much as a atheist, but as a conscious educated intelligent individual, that now has the knowledge available to me to know the formative processes of the Universe, and how they manifest on Earth. So there is no need for a creator God, standing over us. We can know that we and any other 'being' we might meet throughout the Universe are manifestations of movement over a long period of time, and ultimately I am fully responsible for my journey with the Universe. There is no father God defining my existence. There may well be other beings throughout the universe that have been called angels and demigods in the past, but they too are manifestations of movement, and of a similar spiritual nature as ourselves. We may even interact and find purpose in rallying to their cause, just as we may rally to the cause of a human leader, but ultimately it is now me and the Universe.

In many ways this is an image of the Aquarian attitude to life. They stand very firmly in their own knowing, in fear of no authority, blazing the path they feel is correct for them, with a great concern for their community, and humanity at large. So is this the message or the stance for humanity to strive towards for the next 2000 years?

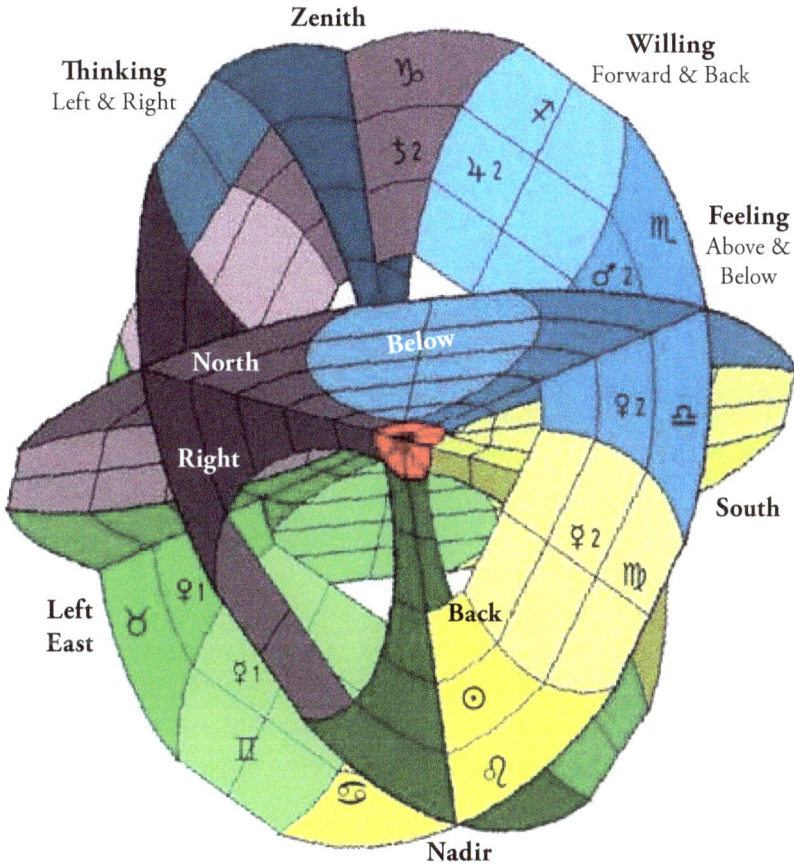

The Universe is a big place, and so I am not suggesting an attitude of rational materialistic atheism, but one of acknowledging the Universe as the wondrous being that it is, and honor all that is there, with the perspective, that we are active and responsible participants, with all that is there. In the past all this was rapted up in religious and mystical jargon based upon faith and belief in something beyond us. It is no longer beyond us. We can know, we do know what is there and what the electro magnetic movements of the planets are doing to us. We can become objective with their influence in their life and see it is not devils or Gods doing this to us, it is us living out the results of our own actions, and thoughts and feelings that we have built into our energetic bodies, and which are played upon by the movements of the planets. Everyone with a internet connection can now know this in great detail and for free. Without this intimate knowing of the planetary effect how can there be an objective sense of self within the Universe. It is always something doing it to you. Yes it is the planets, that are carriers of your fate. It is through becoming aware of their message and passage that we can be truly objective and in no fear of a 'father God'. By learning this Aquarian art you are internalising the 'doing it to you God', and you will have the experience of you are doing it to yourself. This is the next great leap forward for humanity. I hope Astrology will be practiced by almost everyone. It is no longer a science for the elite, it is available for everyone, and it is surprisingly easy, and with computers the sophistication of the information available to us in seconds is way beyond, what most of the best trained Astrologer of antiquity could have even dreamed of.

It has been interesting reading Paracelsus recently, and as great a 16th century intuitive sage as he was, who completely supported the 'As Above So Below' reality of Astrology, it is very clear he did not have the tools of personal Astrology available. He did not KNOW to what degree the planets played upon his psyche, as he could not look, as he could not draw horoscopes. I am in no doubt with the tools available so easily today, several of his understandings would be quite different.

As I write this, I have been reflecting that as I have worked more over the last few years, with magnetic north, the more I have come in my own journey to 'killing the father God'. I wonder how the journey itself, has created the experience, that shows the meaning of this refocus.

Epilogue

from 'Biodynamic Chemistry'

This journey began in 1996 with the help of John Perham, who tutored me in Brookside Soil Science, and Grant Patton who encouraged me with his Biological Agricultural Scientific understandings. In the later stages of this process, Hugh Lovel has been an inspiration with his Quantum view of life, filtered through and beyond the Biodynamic method. Thank you all for the prods and nudges.

I also wish to make a special thanks to the Astrological Vortex, Dr Rudolf Steiner, Dr Rudolf Hauschka and Dr Eugen Kolisko for the work they left behind them, from which this 'idea' has been developed. I trust it was their intention that this sort of endeavor would arise from their great works.

I also thank my wife Caroline Lowry for her generous and seemingly unlimited support of my many endeavors. It is her day to day gifts that have made this book physically possible and a joy to undertake.
1 July 2005

The last 6 years has seen refinements made to the details of my initial work, especially with regards to how the Periodic Table sits with the Earth's electro magnetic field, and the 3D sphere. This edition takes most of this into account. During this period I have also explored the applications of this information to actual 'problems' in both human and agricultural spheres and remain convinced of the value of the information this 3D Periodic Table produces.
30 July 2011

What has been presented here, is the outline of a new methodology that still needs to be worked upon further. I have no doubt it can answer most questions that arise in the process of maintaining health of both people and nature. My aim in 'giving this away' now, is so others can join me in this exploration process, and for humanity as a whole, to have access to the ancient knowledge and techniques, that are essentially our birthright. Here is a safe, free and thorough methodology that can be accessed by anyone with a circle.

I appreciate there is some background understanding needed to really diagnosis problems , and to then identify the solutions. But like all modern sciences, some training in required for something to be used, at its most powerful levels. Anyone can start though by gathering elements you know you need.

My books and range of DVDs are available to help in the learning process. I anticipate there will be seminars and training sessions available in the future as the need arises. Please do make the effort to get the overview I provide as this acts as the rational basis for the practical parts of the method.

While I have a facebook page – "Adopt an Octagon" for the spreading of the word, I also have a yahoo group – 'garudabd@yahoogroups.com' that could provide a more useful forum for discussion of this subject. I do not find facebook particularly functional for ongoing discussion and archiving. So if you are interested in pursuing this subject further join the yahoo group and start asking questions.

My website www.garudabd.co.nz, is where I put this side of my activities, while any purchases are run through the BdMax online shop. www.bdmax. co.nz .

This has been a fantastic journey, which began the moment I left high school. 43 years later this is where it has come to. I feel very privileged to have been on this somewhat individual process of unfoldment, which has lead to the most fundamental of universal truths. A bottle of water, sitting in a circle in the Sun.

I am writing this on the 90th anniversary of Dr Steiner's Agriculture lectures, which has been the doorway into his outstandingly practical worldview. My efforts have been focused on understanding and applying these, classically difficult lectures. A task I believe I have achieved. My book 'Energetic Activities' is where I physically pulled these lectures apart and reordered them, so the two dominant theological discussions in this course, are seen separately. In this bold act I hope a clear theology can be seen by the Biodynamic community, so that a coherent and functional story can be told to the rest of the world. Biodynamics, can not expect to be taken seriously by a scientific world, when every practitioner has a different theology, especially when their story is developed upon the denial of significant sections of its founding document. It is time to move on from this, a clear rationally comprehendible story is available, free to all.

I trust my efforts, both practically, through the effectiveness of the unique products I have created, and through my demonstration of chemistry's coherence with the order Dr Steiner described, provides the Biodynamic community with enough evidence to take my suggestions to their 'unknowns', seriously.

Beyond that community, I trust my efforts provides others with a sequential story of universal creation, and simple techniques by which they can interact with it.

Here's to the future development of humanity within the principles of universal harmony.
4 September 2014

Three more years, and I have had time to look further into the Lanthanides, make more of the diagrams magnetic focused, while experiencing further, the power of the remedies gathered from any circle.

Time and experience continues to prove the validity of the findings presented here. My experiences are being repeated and strengthened by Hugh Lovel, an international agricultural consultant and educator. I have pestered Hugh with this work since 2000, and he has slowly worked his way into it. He recently wrote – "You have my vote for making sense of the periodic table. I'm always following up and confirming your intellectual realizations. My hat's off to you, my friend. I owe you a great (and growing) debt for deeper understanding of biodynamics and beyond, which I hope including mention of your work in the tail end of my book, Quantum Agriculture, is a first payment for. Thank you , thank you, thank you. Best wishes, Hugh Lovel"

I appreciate the practical applications of this approach have some way to go, and a full 'materia medica' will be needed before general use can expected. These will take time. There is enough here now, for the interested person to start. Collect the energies and start experiencing them. There are only four players – Spirit, Astral ,Etheric and Physical. Health is when they are all in their right place. It is that simple, and here is the instructions of how to do it, free of charge and safely. Dr Steiner's medical lectures are online and free to access. Here is an adventure of several lifetimes – Enjoy.

2017

Four years on and another hardcopy printing is required. The larger format allows for larger pictures and font. Along with 4 more years of experience. I am still convinced the suggestions made here have value and well worth the effort it takes to use this material. Others have used the remedies suggested here and report outcomes in line with the suggested effects. Try it for yourself and report back your experiences.

8.9.2021

We hope this new format will make working with this material easier. Good results continue to arrive from all that has been discovered through this approach.

18.12.2024

Bibliography

(1) **Agriculture**, ISBN 0-938-250-37-1, BDFGA USA

(2) **Spiritual Science and Medicine**, Steiner Books, 1920,
ISBN 0- 89345-263-7

(3) **Anthroposophical Spiritual Science and Medical Therapy**.
Rudolf Steiner Archive, April 11 1921 to April 18 1921,
http://wn.rsarchive.org /Lectures

(4) **Spiritual Relations in the Human Organism**, Mercury Press,
Oct 20 1922 – Oct 23 1922

(5) **Fundamentals of Anthroposophical Medicine**, Mercury Press,
Oct 26 – Oct 28 1922

(6) **The Healing Process** 1923-24, Anthroposophical Press
ISBN 0-88010-474-0

(7) **Pastoral Medicine**, Rudolf Steiner Archive,
http://wn.rsarchive.org /Lectures 8th Sep 1924 to 18th Sep 1924

(8) **Biodynamics Decoded**, The Garuda Trust, ISBN 0-474-09003-1

(9) **The Working of the Planets and the Life Processes in Man
and Earth** Dr B. Lievegoed, BDA UK
http://rimu.geek.nz/garuda/Agriculture/Ag%20Course%
20remix%203.pdf

(10) **Biodynamic Questions, Astrological Answers**,
The Garuda Trust. ISBN 978-0-473-28959-1

(11) **The Twelve Groups of Animals**, Dr E Kolisko, Clunies Ross
Press, ISBN 0-906492-06-8

(12) **Hortresearch Bird research document**
http://www.bdmax.co.nz/report2/report2.htm

(13) **Agriculture of Tomorrow**, L Kolisko , Kolisko Archive Publishers,
UK, ISBN0-906492-00-9

(14) **www.garudabd.org**

(15) **Enzo Nastati**
http://www.considera.org/reslit.html
see his commentary on the Agriculture Course.

(16) **The Energetic Activities** – The Garuda Trust,
ISBN 978-0-473-19960-9

(17) **Introducing Anthroposophical Medicine**, Rudolf Steiner,
(especially the lectures of Oct. 26th – 28th, 2022)

(18) **The Nature of Substance: Spirit and Matter**, Rudolf Hauschka,
RUDOLF STEINER PRESS; Revised ed. edition (15 February 2008),
256 pages.

(19) **Circle of Everything**
https://garudabd.org/2019/06/05/the-circle-of-everything/

Wells Cathedral Chapter House
Gyroscopic Periodic Table

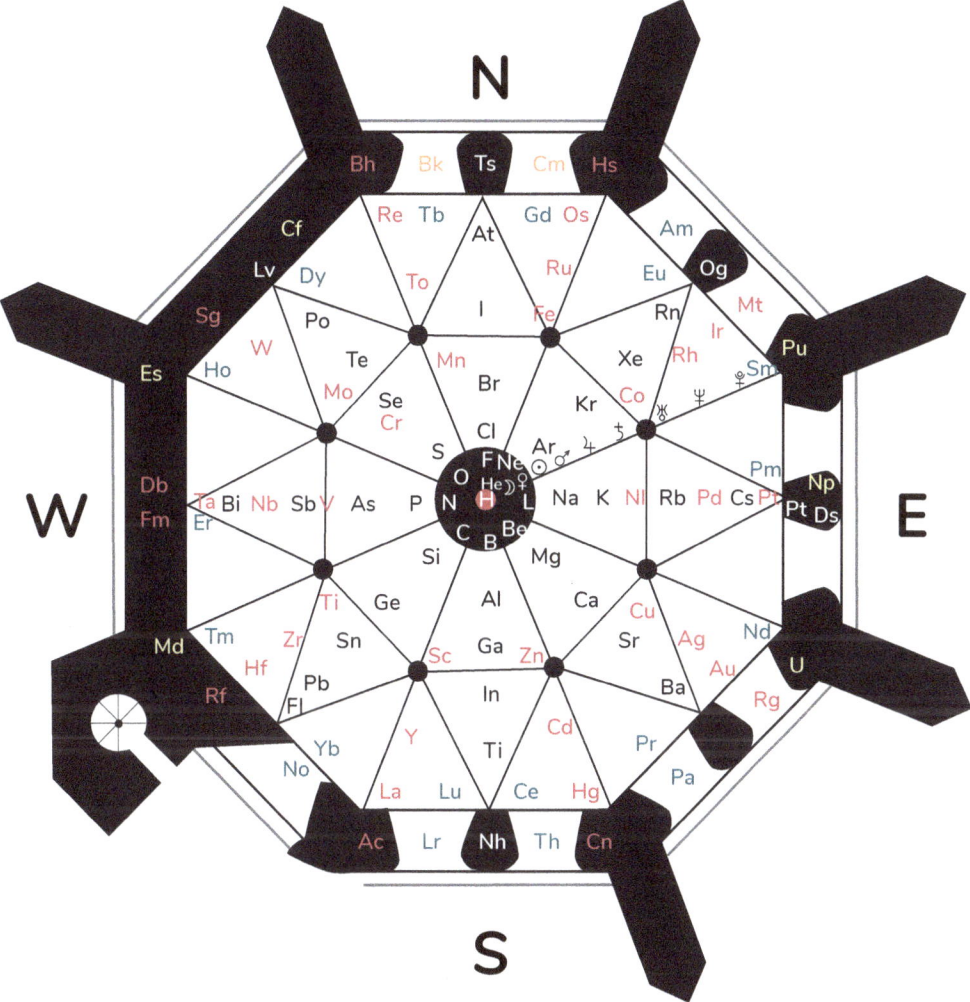